RADIO IN REVOLUTION

THE MEXICAN EXPERIENCE

William H. Beezley, series editor

Radio in Revolution

*Wireless Technology and State
Power in Mexico, 1897–1938*

J. JUSTIN CASTRO

UNIVERSITY OF NEBRASKA PRESS
Lincoln and London

Library of Congress Cataloging-in-Publication
Data

Castro, J. Justin, 1981– author.
Radio in revolution: wireless technology and state
power in Mexico, 1897–1938 / J. Justin Castro.
pages cm. — (The Mexican experience)
Includes bibliographical references and index.
ISBN 978-0-8032-6844-9 (cloth: alk. paper)
ISBN 978-0-8032-8678-8 (pbk.: alk. paper)
ISBN 978-0-8032-8872-0 (epub)
ISBN 978-0-8032-8873-7 (mobi)
ISBN 978-0-8032-8874-4 (pdf)
1. Radio in propaganda—Mexico—History.
2. Radio broadcasting—Political aspects—
Mexico—History. 3. Radio broadcasting
policy—Mexico—History. 4. Mexico—Politics
and government—1867–1910. 5. Mexico—Politics
and government—1910–1946. I. Title. II. Series:
Mexican experience.
HE8697.85.M6C37 2016
320.97209'041—dc23
2015025360

Set in Minion by Westchester Publishing Services

CONTENTS

ILLUSTRATIONS

ACKNOWLEDGMENTS

I am most grateful to my mentors. Terry Rugeley, my PhD adviser at the University of Oklahoma, never wavered in his support for me or this project, which began as a dissertation under his tutelage. He introduced me personally to archives in Mexico City, commented often and efficiently on early drafts, and has become a steadfast colleague and friend. Likewise, I am thankful for Patti Loughlin and Ben Kracht, who were crucial to my development as a scholar during my formative years as an MA student and an undergraduate.

Other professors at OU assisted in my academic development, helped me jump through the hoops of graduate school, and provided excellent suggestions for improving my manuscript. Especially deserving of recognition are Jim Cane-Carrasco, Sterling Evans, Albert Hurtado, Raphael Folsom, Robert Griswold, José Juan Colín, and Alan McPherson.

I also have nothing but gratitude for all my OU compatriots who provided friendship and good conversation, allowed me to bounce ideas off them, read my work, and helped me blow off steam. Most specifically, I thank Gary Moreno, Ariana Quezada, Michele Stephens, Margarita Peraza Rugeley, Doug Miller, Courtney Kennedy, Jeff Fortney, Patti Jo King, Matt Caire-Pérez, Elena Llinas, and Nicolas Gordon.

I finished this manuscript as an assistant professor at Arkansas State University. I have been lucky to have had the most amazing colleagues. The entire Department of History is great. Joe Key, Pam Hronek, and Cherrise Jones Branch especially went out of their way to help me succeed as a scholar and teacher. My dearest friends in A-State's informal College of Humanities and Social Sciences faculty research group were invaluable as friends and reviewers of my work: Enrique Bernales Albites, Vicent Moreno, Jason Barrett-Fox, Kat Lecky, Ryan Lecky, Eric Cave, Jacob Caton, Kristi Costello, Claudio Eduardo Pinto, Michele Merritt, and Carmen Williams. A generous monetary award from A-State's Faculty Research Award Committee helped me put the finishing touches on this monograph. The award allowed me to hire an amazingly talented research assistant, Rachel Widner.

During the last six years of researching this book, I have also benefited from the comradery, advice, and assistance of wonderful people from the United States, Mexico, and Colombia. In particular I would like to thank José Luis Ortiz Garza, Robert Howard Claxton, the late Paul Vanderwood, Jorge M. Rolland C., Marioli Lombrera, Sonia Robles, John A. Britton, Elissa Rashkin, Celeste González de Bustamante, Luis Coronado Guel, Michael Matthews, Elena Jackson Albarrán, Gisela Cramer, Mary Roldán, Amanda Ledwon, Robert Jordan, Andrew Paxman, Andrew Grant Wood, Barry Carr, Diana Montano, and Steven Bunker.

This research would not have been possible without archivists and librarians. I especially want to recognize the staffs of the Archivo General de la Nación de México, Archivo Histórico del Distrito Federal, Archivo Histórico Genaro Estrada de la Secretaría de Relaciones Exteriores, Archivo Histórico Universidad Nacional Autónoma de México, Centro de Estudios de Historia de México, Fideicomiso Archivos Plutarco Elías Calles y Fernando Torreblanca, Universidad Iberoamericana—Acervos Históricos, Wisconsin Historical Society, Hermoteca Nacional de México, and the OU and A-State libraries.

The University of Nebraska Press was amazing to work with. Bridget Barry possessed all the qualities of an excellent editor. She was supportive, efficient, and honest. Emily Wendell was also immensely helpful. The critiques of the anonymous readers improved this manuscript greatly.

Of course, I owe a great debt to my family and friends who have been my bedrock of love and support over the years. Thank you Angela Castro, Olivia Castro, David Castro, Lillian Castro, dad. Mom, I wish you were still alive to see this book come out, perhaps upon a distant star we'll sit and chat about it.

RADIO IN REVOLUTION

Introduction

A Tale of Two Revolutions

On Valentine's Day, 1924, the residents of Mexico City awoke to discover a strange and tragic story in the daily *El Demócrata*. After scratching a brief note into a piece of maguey, a middle-aged man purportedly distraught over his wife's adulterous affair had climbed to the top of one of the radio towers near Chapultepec Castle and jumped to his death. The article described his end in gruesome detail. The next day, however, the tale took a dramatic turn when the newspaper revealed that instead of a heartbroken lover, the deceased had actually been a member of a rebellion recently started by Adolfo de la Huerta, a former high-level member of the revolutionary leadership who was attempting to overthrow the president, General Álvaro Obregón Salido (1920–24). According to this new twist, the death in question had actually resulted when the man electrocuted himself in a botched attempt to sabotage an important component of Obregón's communications system.[1]

This tale of espionage, sensational as it was, touched upon real fears. Just as surely as infidelity wreaked havoc on Mexico City marriages, Delahuertistas had taken a number of important wireless stations in the country. Spies were operating radios from the rooftops of crowded houses in the nation's capital. General of the Federal District Arnulfo Gómez, then ruthlessly hunting down insurrectionists, went so far as to tell journalists that radio was "the principal enemy of the government."[2] He demanded that radio owners register all their devices. In reality, Gómez was articulating an uneasiness about wireless communications

FIG. 1. "Desde la cúspide de una torre inalámbrica del bosque de Chapultepec se arrojó un sentimental y excéntrico suicida," *El Demócrata*, February 14, 1924. Courtesy of Archivo General de la Nación de México.

that had developed among military commanders during the preceding Mexican Revolution (1910–20). The Revolution was violent and destructive. Erupting during the overthrow of the Porfirio Díaz administration (1876–80, 1884–1911), the Revolution also sired a number of subsequent revolts, of which the De la Huerta Rebellion was arguably the most significant.

During radio's first forty years of existence in Mexico, political leaders often perceived radio as a means of control and rebellion. Porfirian generals close to Díaz had first brought radiotelegraphs to Mexico in order to expand state power over the nation's little-controlled frontiers. The Revolution briefly disrupted this plan while expanding the use of the technology as a tool of warfare. Whatever the situation, radio remained connected to political and military authority. Just as the Díaz administration hoped to unite the nation's hinterlands to Mexico City, the leaders of the various revolutionary factions used the medium to gain dominance over their enemies. Territorial oversight and communications went hand in hand because to have the former, one had to develop and to maintain the latter.

Revolutionary leaders turned to radio, a relatively new technology, to eliminate the difficulties of preserving and expanding wire communications. Telegraphs and telephones were the dominant forms of electronic communication in the 1910s, but they were easily disrupted. All one had to do was cut, at any point, a part of the miles of indefensible cables, and the connection failed. Radio stations had no interconnecting wires, making them harder to sabotage. As a result, radio gained a more prominent place in communications.

Before the end of the Revolution, wireless technology went through a revolution of its own. American engineers and experimenters had made the adjustments necessary to transform radio transmitters and receivers from Morse-code, point-to-point communications tools to voice-broadcasting devices that could reach thousands of people. Consequently, commercial broadcasters in the United States began airing music and advertisements. Almost immediately thereafter, Mexican engineering students, businessmen, and military communications specialists started their own experiments.

Following the introduction of broadcasting, Revolution-era leaders used the new medium to spread revolutionary culture. Although the military used radiotelegraphy to help its soldiers contain rebellions, and rebels used it to fight the government, political leaders used broadcasting to spread their ideals among labor and peasant groups, foreign audiences, and a small-but-growing urban middle class. This transformation had huge ramifications. Radio was no longer used solely as a point-to-point tool of the postal service, seafarers, and the military; it had become the first electronic mass medium, a crucial tool for integrating the general population across the national territory by means of generating consent.

The development of radio and its subsequent radical change during the revolutionary era beg a number of questions. Did wireless technology affect the outcome of the Revolution and the revolutionary state? How did the extreme violence and instability affect radio development in Mexico? How did the subsequent government leaders, whose control over the state was still challenged, attempt to use and manage radio? To what extent could they? All of these questions, in turn, relate to more fundamental issues, namely, the complex role of electronic communications in revolutions, government, and modern society.

This book presents three main arguments. First, that radio technologies were crucial to certain attempts to centralize state power in Mexico during the late Porfiriato. Second, that radio was decisive to the outcome of the Revolution and subsequent plans for solidifying the nation-state. And third, that rebellions during the 1920s pushed government leaders to pursue more authoritarian radio policies and practices that, in turn, helped them consolidate their control.

Ultimately, radio became a pillar of the single-party state that ruled from 1929 to 2000. This book examines the foundations of this relationship between electronic wireless technologies (radiotelegraphy and radio broadcasting), single-party rule, and authoritarian practices in Mexico. Although this study discusses Mexico and radio specifically, it also contributes to larger discussions about the interplay of communications technologies, revolutions, governance, democracy, and authoritarianism.

Radio, of course, was not the only technology transforming Mexico. It was but one of a number of important innovations that occurred during the long administration of Porfirio Díaz and the subsequent revolutionary era. Fueled by industrialization and scientific research in Europe and the United States, a massive technological transformation was sweeping across much of the inhabited world. Steamships, railroads, electricity, machine guns, and telegraphs all proved useful to agents of imperial powers who were rapidly expanding global trade and exploiting colonial possessions. They also proved useful to the Díaz government.

Similar to European and U.S. empire-building practices, Díaz's communications and military leaders first adapted radio as a means to advance trade and connect frontier hinterlands. In the decades preceding his rule, especially before 1867, civil war, banditry, and foreign invasions had all wreaked havoc on Mexico's transportation and communication networks. The government had failed to maintain roads adequately, and in some ways the region was more disunited than it had been under Spanish rule before 1821. In the 1830s and 1840s, Mexico lost nearly half of its territory—a huge portion of the northern periphery with which the capital had possessed weak transportation and commercial links—to the hungrily expanding United States. To the south, creoles in the mostly disconnected Yucatán briefly declared independence and became absorbed in a bloody civil war with Mayan communities upset with the costs of maintaining the nascent Yucatecan state and the associated growth of political violence. Between these extremes, even many of the more central states maintained a relatively high degree of isolation from federal authority. Members of Díaz's military and political leadership viewed increasing communications as essential to establishing greater state control.

Attempts at advancing communications had been under way since the 1850s and included such innovations as the first railroads and telegraph lines. The invasion of French forces in 1862 temporarily halted progress in creating these links. Progress resumed, however, following the French victory. France's ruler, Napoleon III, made Archduke Maximilian of Hapsburg the Emperor of Mexico in 1864. During Maximilian's reign, and following his defeat and execution by nationalists under the leadership of Benito Juárez, laborers slowly added new railroad tracks and telegraph lines. The Veracruz–Mexico City railway, for example, reached completion in 1872.[3]

It was not until the Díaz era, however, that modernization advanced at a consistent pace. With strong support from Díaz, the newly created Geographic Exploration Commission worked to systematically survey and map Mexico. Obtaining a firmer understanding of the landscapes, people, and resources of Mexico was essential to consolidating power, defeating rebellions, and creating a stable environment for economic investment.[4] The government funded and directed the maintenance and construction of a large number of roads, and the Secretary of Communications and Public Works (SCOP), in collaboration with foreign capitalists, expanded railroads at an astounding pace. There were less than four hundred miles of railroad track in 1877; by 1910, there were over eleven thousand miles of track.[5]

Meanwhile, cable telegraphy experienced similar growth. Railroads and wire communications went hand in hand. Train stations often doubled as telegraph offices. Juan de Granje, a Mexican merchant of Spanish origins, introduced telegraphy to Mexico in 1849. As with the first railroad, the construction of the country's initial telegraph line occurred in the 1850s between the port of Veracruz and Mexico City. It was not until the late 1870s, however, that cables connected the Atlantic and Pacific coasts. Over the next two decades, the newly created Department of National Telegraphs (DGTN), under the direction of the SCOP, started expanding lines, once again in collaboration with foreign businesses, but also the military, to the country's fringe territories.[6] However, most of the country still remained unconnected. Particularly problematic were Mexico's two peninsulas, which remained separated from Mexico City by harsh terrains and bodies of water. The SCOP struggled, often unsuccessfully, to construct telegraph lines through the jungles and deserts and to lay them under the Sea of Cortez. These areas are where the Mexican government initially focused its radio projects during the first decade of the twentieth century. Following the Revolution, the peninsulas continued as a focal point for state radio operations.

These developmental processes, of which radio was a prominent component during the twentieth century, were, in the words of political theorist James C. Scott, ways in which state officials made "a society legible, to arrange the population in ways that simplified the classic state functions of taxation, conscription, and the prevention of rebellion."[7]

As in empires that managed widespread territories and diverse cultures, Mexican political leaders relied on information just as much as the military to obtain greater control over the people they wished to influence. Information, control, and violence went hand in hand. Historian Raymond B. Craig highlights this very point in his work on Mexican cartography, arguing that "cartography and the military, knowledge and force, were mutually reinforcing."[8] Just like maps, early radio technology was a tool of state centralization and military force, not just during the Porfiriato, but also during the subsequent revolutionary era.

Most publications on Mexican radio focus on the 1930s through 1950s, though some works discuss later periods as well.[9] This book delves deeper into the roots of Mexican radio.[10] Its examination of the technology's origins in Europe and the role of Mexican ambassadors in obtaining wireless equipment broadens the historical and geographical context of Mexican radio studies. This approach has also enabled this work to bridge the historiographical divide between works on communications and empire building and those on radio and Mexican nationalism and state formation.[11] The processes of state consolidation in Latin America and the construction of empires by the United States and European powers were closely related, if not the same. In addition, this study places a much stronger emphasis on the role of the Revolution and subsequent violence on state broadcasting policies than previous studies. It was a fear of continued insurrection, together with the drive to consolidate control over the state, that pushed the military to expand its wireless operations and national political leaders to later increase government broadcasting, to collaborate with specific commercial stations on programming, and to limit the scope of oppositional political speech over the air. The Revolution is why populist politics emerged over the airwaves in Mexico earlier than in other Latin American nations.[12]

In addition to its contribution to studies of media and technology, this work also enriches the existing literature on the Porfirio Díaz era and the Mexican Revolution. The publications covering these two periods are so numerous that listing them all would be a cumbersome task.[13] Some monographs about the Porfiriato discuss technology, public works projects, and modernization, but none excepting a handful of rare and rarely read works on telecommunications discuss the importance of wire and wireless communications to Porfirian schemes.[14] Similarly, most books on the Revolution and the subsequent governments tend to focus on ideology and culture more than on technology and structural development. Some of the most significant works on the Revolution make reference to technologies, but none emphasize the point.[15]

It should not be surprising, however, that radio was important in the Revolution, first as a military tool and then in the 1920s and 1930s as a mass medium to spread revolutionary culture. There has been a long-standing relationship between communications and revolution. Often scurrilous penny papers helped build momentum for the French Revolution.[16] Following the Mexican Revolution, radio became an essential component of revolutions and counterrevolutions in Cuba, Guatemala, and Nicaragua. And in the early 2010s, scholars and journalists have debated the importance of Internet social media sites in the Arab Spring.[17]

The Mexican Revolution provides the earliest example of radio technology in a major social revolution. It shows that electronic communications have been important to both challenging and maintaining government control since their introduction. Radio during the Revolution was a novel but valuable tool for administrating government in conquered territories, organizing military operations, and conducting foreign relations.

Unlike penny presses, or Facebook for that matter, radio was initially a point-to-point form of communication, not, at least intentionally, a mass medium. Its value lay in its supposed privacy. Thus it is more difficult to find transcripts of what people said via radiotelegraphy. Existing reports are often from individuals who spied on other people's conversations. Still, radiotelegraphy's value as a communications tool should not be underemphasized. Messages via radiotelegraphy, in the words of technology historian Daniel R. Headrick, were often more directly "relevant to the conduct of government, to the outcome of wars, to the pursuit of business, even to the lives of ordinary people."[18] What people said via radiotelegraphy was often crucial to business or military strategy, or, if personal, extremely intimate. The use of radiotelegraphy during the Revolution also had important ramifications on the subsequent broadcasting era. Specialists who operated wireless telegraphs became immensely influential in broadcasting operations.

Chapter 1 discusses how the Díaz administration obtained radio equipment from Europe, the influence of European imperial practices, and why Mexican state officials decided to build the first radiotelegraph towers in their country's peninsulas. Before the Revolution, the Díaz government mostly used radio for these frontier-connecting developmental programs and for maritime trade, though by 1907 military experiments with portable wireless devises were also under way. Radio was a tool of business and of state communications, centralization, nation-state building, and military advancement. The Díaz government used radiotelegraphy to aid the process of military domination and colonization in these frontier provinces that former administrations had failed to fully incorporate. This process mirrored similar strategies used by the empires of Europe and the United States in their colonial schemes. The Díaz administration's attempt to build a state-owned national radiotelegraphy network also provides an example of the government's increasingly nationalist bent during the decade preceding the Revolution.

Radio operations by state officials and insurrectionists increased during the Revolution, an important development that is overlooked by most scholars but addressed in chapters 2 and 3 of this work. Even beyond radio, there is a general absence of publications on the role of modern technology in the Mexican Revolution. This is a strange void, considering that there are enough books about this great upheaval to fill a small library and that the Revolution was unquestionably a testing ground for modern machines. The first bombings of warships from an aircraft occurred during the Revolution.[19] Alongside machetes and old rifles, modern technology in the form of airplanes, cars, trucks, machine guns, trench warfare, steam-powered and electrified ships of war, and, of course, radio, all contributed to the armies of the various fighting factions.

Of these technologies, radio has been especially ignored despite the DGTN's construction of ten stations by the time the forces of Francisco I. Madero (a wealthy landowner from Coahuila) overthrew the octogenarian dictator. Although briefly interrupted by the subsequent revolution turned civil war, political and communications leaders continued to expand on the practices and ideas of their Porfirian predecessors. In fact, many of the SCOP and DGTN employees who built government radio stations in the 1910s and 1920s had either worked for the same agencies during the Porfiriato or were trained in Porfirian universities. Radio also proved crucial to important battles during the Revolution. Each faction possessed its own highly valued wireless telegraphy operators.

The Constitutionalists, who were the victors of the Revolution under the leadership of Venustiano Carranza (1915–20), relied on these specialists, increasing their ranks while reconsolidating control over larger and larger portions of the country. Because of continuing instability and the threat of anti-Constitutionalist communications, Carranza decreed state monopolization of wireless technology, which legislators codified in the Constitution of 1917. When General Álvaro Obregón took the presidency in 1920, following the last successful coup of the Revolution, twenty-seven radiotelegraph stations operated across the country in addition to a number of portable military radios.[20] During the Revolution, radio was crucial to the military success of the Constitutionalist forces and to their simultaneous attempts at nation-state building.

In 1921 the first broadcasting exhibitions took place during the Mexico City centennial celebrations of independence from Spain. Broadcasting dramatically changed radio. As discussed in chapter 4, this new use of radio technology revolutionized politics as the Obregón and the subsequent Plutarco Elías Calles administration (1924–28) obtained a firmer, if still contested, control over the state. Whereas military radio helped the army contain rebellions, broadcasting allowed revolutionary leaders to spread their ideals among labor and peasant groups, foreign audiences, and a growing urban middle class.

Unlike Carranza, subsequent leaders, beginning with Obregón, abandoned the idea of having the state monopolize radio. Recognizing the economic and political potential of broadcasting, Obregón followed the advice of state technicians and representatives of a prominent Mexico City radio club that promoted private commercial and experimental radio use. The relationship forged between commercial radio leaders—most notably the Azcárraga family and El Buen Tono—and government officials during the 1920s and 1930s (chapters 4 and 6), established a mutually beneficial partnership that helped solidify single-party rule; commercial broadcasting under a handful of stations, including XEB, XEW, and XEQ; and a vibrant and complicated Mexican nationalism.

The last half of this book addresses the increasingly diverse uses of radio during the mid-1920s and 1930s. Using a thematic approach to discuss radio in its varied applications, the last two chapters both focus on the presidency of Plutarco Elías Calles, the Maximato (1928–35: A period when Calles wielded a strong influence over three short-term presidents), and the first four years of the Lázaro Cárdenas presidency (1934–40). Chapter 5 continues the discussion of radio in military and infrastructural endeavors, while chapter 6 elaborates on the huge cultural and political changes brought about by broadcasting. Both chapters emphasize the Calles presidency and the Maximato over the more studied Cárdenas era because they have received substantially less investigation, and because it was during the decade prior to Cárdenas's election that broadcasting became an important political tool.

Chapter 5 examines the role of radio in the military, the suppression of rebellions, infrastructural projects, and foreign relations. It also explores radio legislation, which was heavily intertwined with the insurrections of the late 1920s and relations with the United States, Central America, and the Caribbean. During the Calles and Maximato years, radio became an even larger component of state control, military professionalization, and the modernization of transportation networks.

Radio also became an essential part of revolutionary populist politics during the Calles and Cárdenas eras (chapter 6). This chapter points out that by the time that Cárdenas took office, presidents had been using broadcasting for a decade and wireless telegraphy for thirty-five years. The traditions of using radio to relay inaugural addresses, congressional speeches, state-sponsored events, and presidential campaigns had all been placed upon sturdy foundations built by Calles and, starting in 1929, the political party he helped found, the National Revolutionary Party (PNR). Cárdenas's ability to rouse large portions of the nation and to increase government corporatism had as much to do with the maturation of populist politics, the broadcasting industry, and radio consumerism as with his own special skills as a political leader.

In its totality, this book examines the advancement of radio communications in a time of extreme technological, political, and societal change. It focuses on the period between the establishment of a Mexican radio system under Porfirio Díaz at the turn of the twentieth century to 1938, when President Lázaro Cárdenas famously used broadcasting to rally the Mexican people during the expropriation of foreign oil companies—a high point in state broadcasting. It was also in 1938 that the last serious rebellion stemming from the Revolution against the central government took place and the greatest number of government stations were in operation. Thereafter, state leaders began to dismantle government broadcasting stations and became increasingly cooperative with corporations and the U.S. government. The Revolution became less revolutionary. Instead of treating the Porfirian and revolutionary periods separately, or dividing radiotelegraphy from broadcasting, this work places them together in order to better trace political and technological continuity and change. Radio development, like beliefs in nationalism and the progressive nature of technology and modernity, bridged the positivist and revolutionary eras.

The most significant consequence of the Revolution on radio development was that military concerns shaped how government leaders, mostly military men themselves, perceived the medium.[21] A strong focus on preventing radio from aiding new rebellions or being used to spread antigovernment messaging limited the brief democratic tendency in radio that occurred with the advent of broadcasting and the coinciding growth of private radio under Obregón. As a result, the state continued to play a strong role in wireless use and growth. Those whom the government allowed to transmit radio services faced strong regulatory restrictions, and all political broadcasting became dominated by political leaders who collaborated in the PNR and its successor parties, the Party of the Mexican Revolution (PRM, 1938–46) and the Institutional Revolutionary Party (PRI, 1946–present).[22] Hence, radio became crucial to the solidification of political power in the hands of a party that would go on to rule for over seventy years.

Born from experiments conducted in laboratories in Europe, radio became a crucial tool for the acquisition and solidification of state power in Mexico. From the moment that the first devices reached Mexico in 1899, they were a tool of state expansion and nation building. But as the Revolution clearly demonstrated, the technology, when not monopolized or firmly controlled, could also be a force of rebellion, resistance, and factionalism. As radio was developed and amplified during one of Mexico's most massive struggles over state power, controlling it became an essential component of war and governance.

1   Porfirian Radio, Imperial Designs, and the Mexican Nation

The effective government of large areas depends to a very important extent on the efficiency of communications.

Harold Innis, *Empire and Communications*

Staring out over the gulf waters off the coast of Veracruz, young telegrapher Alejandro Gutiérrez dreamed of accomplishing great things. It was 1902 and the government had selected him to head experiments with German radiotelegraphy equipment. Gutiérrez had once again successfully signaled a fellow radio operator on the steamship *Melchor Ocampo,* a coast guard vessel. He achieved a clear reception at distances of seventy miles "under bad and good atmospheric conditions."[1] Although deservedly proud of his accomplishment, Gutiérrez and other communications officials had reached a plateau in their experiments, which had been under way for two years. He could not break the seventy-mile mark. Undaunted, he aspired to connect "Veracruz, Frontera, Progreso, Campeche, Tampico, and Túxpan, for the service of the Public Administration."[2] He believed in the future of the technology, in its ability to expand trade, increase the reach of government, and to facilitate Mexico's rise into the membership of modern nations.

European practices especially influenced Mexican officials. They desired acceptance among the leaders of more powerful nation-states and the power they possessed. Europe's scientists and engineers developed amazing machines that lined the pockets of businessmen and rulers with wealth extracted from far-flung colonies and rising nations like Mexico. Radio was one of the newest innovations. The navies and armies of Britain, Italy, Germany, France, and Belgium raced to incorporate the technology. Political and military agents worked diligently to connect offshore islands and remote resources to metropolitan centers. In the process, they defined world protocol. Mexican officials had similar ambitions, albeit more for gaining control over their own politically defined territory and proving they were worthy of sitting at the international policy table.

The administration of President Porfirio Díaz had made progress toward its goals in establishing a radio network. It had just presided over the groundbreaking of the first public radio stations, one on each side of the Gulf of California. Four years later, in 1906, Mexican representatives participated in an important international conference on wireless technology in Berlin, Germany, aligning with the host country against the British Empire.[3] Germany provided a better deal on radio transmitters and receivers, and unlike their British competitor, allowed Mexicans to operate their own radio systems in addition to owning the equipment. In exchange, Díaz's envoys provided loyalty in diplomatic debates over communication networks. Gutiérrez, the telegrapher with a very personal interest in radio, was a part of something much bigger than his goal of beating the seventy-mile-transmission mark, or even connecting Mexico's gulf ports. He was at the center of a grand struggle between global empires and a Mexican government attempting to use these imperial conflicts for its own nationalist endeavors.

In power since 1876, excepting a four-year stint from 1880 to 1884, Díaz brought relative stability, economic growth, and modernization to much of urban Mexico. Leaders of industrialized empires applauded him for his firm (and sometimes ruthless) rule, especially his protection of foreign investments. Although described by later anti-Porfirista historians as a sellout to foreigners, the Díaz administration had nationalist ambitions. Díaz's advisers understood that technological development was crucial to centralizing power and that the only way to build vast miles of railroads, telegraph lines, and waterworks was to invite in outside specialists and capitalists. Although foreigners often made the greatest profits, the Díaz administration argued that the infrastructural improvements would allow for the growth of domestic capital and consumerism. Meanwhile, the government built new professional colleges, including the National School for Engineers and the National Agrarian and Veterinary School, to train a new generation of Mexican professionals to wean the country from outside influences. The Díaz administration believed it was through infrastructural development and education that a stronger national culture would be born via greater economic and personal exchange, and by force, too, if necessary.[4]

Despite some success in educating Mexican specialists and building a stronger domestic economy, aggressive U.S. economic and military actions created anxiety within the Díaz government by the end of the 1890s. U.S. influence had grown too strong. The Spanish-American War (1898) clearly demonstrated the United States' willingness to impose its will with little regard to the sovereignty of Latin American countries. By 1898, U.S. corporations controlled many of Mexico's infrastructural services, including railways and much of the telegraph system. While aspiring to maintain a workable relationship with the United States, Díaz officials feared U.S. domination and strove to counter American influence by expanding European investments and, by the early 1900s, moving toward increased nationalization of the railways. It was under these circumstances that radio development occurred.

In many ways, the history of Porfirian radio is the story of how technology, competing empires, and rising nation-states intertwined. Mexican consulates in Europe and the United States relayed information back home on new, useful technologies. Governments and businesses of industrialized empires benefited from these transfers, often gaining access to cheap resources and at least partial control over how and where new technologies were used. Imperial governments competed for these relationships with developing nations, including Mexico, to form international and political alliances, expand markets for manufactured

FIG. 2. Porfirio Díaz, 1907. Courtesy of Library of Congress, Prints and Photographs Division, Reproduction No. LC-DIG-ggbain-03727.

goods, and to obtain greater access to natural resources. At the same time, the Díaz administration worked to play growing international competition to its advantage.

In the case of radio, wireless technology developed faster in Europe than in the United States; the largest exporters were in Britain, Germany, and France. After exploring their options and buying some experimental equipment from France, Mexican SCOP and military officials ultimately allied with German companies because they were willing to sell equipment to Mexico and allow Mexicans to operate the services. German agents additionally agreed to help build stations and to assist in the training of radio specialists. The German emperor Wilhelm II and the electric company Allgemeine Elektricitäts-Gesellschaft (AEG) did this to undermine Britain's Wireless Telegraph and Signal Company, which demanded from its customers control over the service in addition to money for the equipment.[5]

The Wireless Telegraph and Signal Company was the first powerful multinational radio corporation, and they were attempting to globally monopolize radio products and services. This power play outraged leaders of other European empires who already resented Britain's domination of undersea cable communications. The company's policies were not appealing to Mexican officials either. During the late nineteenth century, the Díaz administration had worked to increase the number of domestic engineering experts, moving toward a more nationalist developmental plan. Understanding the animosity against Britain, Díaz's communications leaders partnered with the Germans, allowing the government to operate its own radio services and strengthening Mexico's position as an independent power.

With the German equipment, the Mexican government established itself more directly in Mexico's communication systems while using the devices to further consolidate the nation-state. Mexico possessed no provinces outside its own claimed political boundaries, but it did have populations in frontier territories that were resisting central control or were threatened by U.S. usurpation. With the assistance of German engineers, Mexican communications officials designed the first large-scale radio projects in Mexico, building tighter links between Mexico City and these largely autonomous and rebellious regions. Although some private foreign and domestic enterprises used radio in Mexico during this period (often without government permission), state officials drove development. This government initiative exhibits the complicated but intertwined forces of modernization, nation building, empire, and globalization that were under way during the first decade of the twentieth century. It also set the precedent for a strong state presence in radio that would color how government leaders would perceive the medium until the late 1930s. During the Porfiriato the state maintained the strongest control over wireless communications, thus providing evidence that government officials were becoming increasingly nationalist in infrastructural development well before the Mexican Revolution ousted Porfirio Díaz in 1911.

Learning of Radio: Mexican Ambassadors Abroad

The first challenge facing authorities was how to adapt radio, a foreign technology, to their own designs. Embassy officials in Europe first brought news of radio to Mexico. The military shortly thereafter began fact-finding missions abroad to examine and acquire wireless devices. The accounts of these envoys reveal the international breadth and high demand of the technology, the motives behind the businesses and governments first involved with wireless, and the reasoning behind the Mexican government's decision to purchase its own apparatuses.

Interestingly, the story of Mexican radio starts in Italy: a fitting location actually, because it was the birth place of Guglielmo Marconi, the man who first made radio a commercial success. During a business trip to Italy in 1897, Italian royalty and scientists alike showered Marconi with adulation. Publishers spread word of his marvelous innovations. This acclaim was a welcome change for Marconi. The Italian government had initially shown little interest in his inventions.

After painful rejections from Italian officials in 1895, Marconi, with the assistance of his well-connected mother, Annie Jameson—a descendent of the wealthy Scots-Irish family known for its whiskey—had found more success in London in 1896. There, he gained the support of William Preece, the chief engineer of the British Post Office. The British navy also took a strong interest in Marconi's work. In Britain, Marconi established the world's first patent on radiotelegraphy and started the Wireless Telegraph and Signal Company.[6]

Marconi returned to Rome from England to prove his worth and make a profit. He conducted experiments in the palace of the minister of the navy and between the shore of Spezia and Italian ships of war. Marconi successfully communicated with the ironclad *San Martino* at a distance of twelve miles.[7] Political and military leaders, along with the rest of the nation's newspaper readers, followed the trials with "the greatest interest."[8]

The widespread publicity about radio also reached the ears of G. A. Esteva, a Mexican consul in Rome. Recognizing the importance of the technology and the attention it was rapidly receiving, Esteva relayed information about the new marvel "that interests the entire civilized world" to Ignacio Mariscol, the secretary of foreign relations in Mexico.[9] Esteva expounded on his initial statements by sending back publications and more elaborated opinions on the medium's development. His first package consisted of an article in the Italian journal *L'elettricista* by the prestigious professor Angelo Banti, which summarized some of Marconi's work.[10] Mariscol, in turn, disseminated this news to other sectors of the administration, including President Porfirio Díaz, Secretary of War and Marine Bernardo Reyes, Secretary of Communications and Public Works Francisco Z. Mena, and Director of National Telegraphs Camilo González.

Consuming information from across the seas, Mexican officials gathered opinions about radio while seeking out possible purchases. They searched for equipment in countries leading the way in wireless experimentation—Britain, France, Germany, Belgium, and the United States.[11] As the home of Marconi's main business, Britain quickly stood out as a rational place to turn. In November 1898, Adolfo Brule, a Mexican ambassador in London, began sending Mariscol, Mena, and González documents on "the new 'Marconi' system of wireless telegraphy."[12] However, conflicts over who would control the service and cost quickly soured negotiations.

Reports from J. Beuif, a representative in Belgium, dramatically influenced Mexican officials. Experimental radio trials had just started in Mexico (communications officials had bought a couple experimental devices from France), and Beuif's numerous letters about radio tests and negotiations influenced government plans for the medium. The ambassador, much like the Belgian royal court, was particularly intrigued by radio's military applications.

The king of Belgium, Leopold II, had been engaged in a "philanthropic" colonization project in the African Congo. Although, according to Beuif, Belgium was a country "not disposed towards war," Leopold's imperial endeavors—even if paraded as a project of goodwill—were built on exploitation and genocide, a violent imperialist endeavor based on ideas of racial and cultural superiority.[13] Like many other European "civilizing" and economic projects of the era, this mission was also about modernization. It incorporated newly developed technologies including railroads, machine guns, and electronic communications. These tools made possible the suppression of colonial populations while extracting ivory, rubber, and other valuable resources.[14]

King Leopold II, whose greed had generated an interest in matters of geography, economics, and technology, quickly perceived the potential of Marconi's "ethereal" telegraph.[15] In April 1900, while the monarch's forces were suppressing a prolonged rebellion in the Congo, Leopold invited Marconi to the Royal Palace in Brussels to exhibit his new wares. In addition to foreign ambassadors, including the Mexican envoy, the guests largely included military engineers. Befriending the

Belgium general who headed the telegraph division, Beuif collected data on the applications of radio in their armed forces. According to him, Mexico too possessed "talented Military engineers and telegraphers that could take advantage of the new referred to advancements."[16] Reyes and Mena agreed.

The SCOP had obtained its first experimental radios from France in 1899, but Reyes wasted no time in obtaining more equipment. He desired to use wireless technology to increase the power of military expeditions in places including southwest Yucatán, where Maya communities continued to resist the Mexican state. In 1901 he sent a small delegation to Europe under the leadership of Colonel Ignacio Altamira to seek out apparatuses. Over the next five years, Altamira toured Europe and the United States searching out new information. Reyes himself became more involved in obtaining equipment. In fact, he played an important role in sealing the partnership between the Díaz government and AEG, which supplied the equipment and expertise for Mexico's first radio stations. The partnership between AEG (and its successor Telefunken) and the Mexican government would remain strong for two decades, outliving the downfall of Díaz in 1911.[17]

Starting in 1904, word of U.S.-manufactured wireless equipment became more prevalent as the U.S. Navy realized the potential of the technology. Colonel Altamira reported on American radios able to transmit successfully at a distance of fifty miles.[18] The following year, Andrew Plecher, a U.S. businessman, approached the Mexican consulate in Los Angeles with a proposal to establish a wireless system between the state of Sinaloa and the Southern District of the Territory of Baja California.[19] The consul dutifully relayed the information, though the Díaz administration had no desire for a U.S.-controlled radio system, especially in that part of the country where American influences were particularly strong. The Díaz regime was selective in its providers of radio devices, and it was Mexican ambassadors and military officials abroad who provided the administration with information on what was available and how other governments and businesses used the technology.

The First Experiments

The Porfirian government prided itself on the rate of material development under its rule. Continuing the trend, Mena received the country's first wireless devices in the summer of 1899, one of a number of new electric technologies arriving in Mexico.[20] Electricity, still fairly novel itself, had become more prevalent in the largest population centers. Electric lines crisscrossed the air above important thoroughfares. From 1887 to 1911, more than one hundred light and power companies registered in Mexico.[21] Telephone service had expanded too but was still relatively rare, especially in rural areas. The French Lumière brothers had brought cinema to the country in 1896. Shortly thereafter they began filming Mexican movies.[22] Along with phonographs, typewriters, and railroads, radio added to the modernizing landscape (and soundscape).

Elite members of urban society usually saw these technological developments in a positive light. A number of journals and magazines praised the increased mileage of railroad tracks, telegraph and electric lines, and associated goods. The editors of *El Mundo*, *El Mundo Semanario Ilustrado*, *La Revista Moderna*, and *El Cronista de México* published articles relating material progress to the establishment and maintenance of peace and stability, even while railroad development was causing strife in the countryside for many people. Yet criticism existed, mostly from ardent nationalists who resented the growing influence and presence of foreigners, especially Americans. Many Porfirian intellectuals and government leaders, however, believed that exhibiting Mexico's modernity in grandiose public displays at home and in international exhibitions abroad helped increase the nation's status and, in turn, the likelihood of foreign investment, which would be beneficial.[23]

Popular perspectives tended to be more critical and reactive. Some illustrations, including a number of images by the profuse artist José Guadalupe Posada, and coinciding writings by Antonio Vanegas Arroyo often cynically joked about middle-class and elite urbanites who took up bicycling, seeing it as a foreign and passing phase. Posada's work also explicitly demonstrates the fear and critical beliefs circulating about electric trolleys, which started operating in Mexico City in 1900. There was also a fear of electricity itself. Many people held suspicions about the mysterious force moving streetcars, "the invisible colossus that powers them, which shows itself only with sparks and crackles."[24] Similar concerns developed among many who first encountered radio. Not understanding the science behind it, many feared the devil's work. Of course, the trolleys' record of having run over hundreds of people in their first years of operation did not relieve criticism. But as the years passed, electricity became a generally accepted norm in Mexico City, if much less so in the countryside.[25]

The rise of electricity had other more subtle, but important, material effects as well. The dramatic increase in electricity in the 1890s, especially in more industrialized nations, allowed for more efficient production of a wide array of substances and manufactured goods. In the United States, factory workers used electricity to separate chlorine and sodium from salt and brine. Manufacturers used the chlorine, combined with lime, to make bleaching powder for paper and textiles, advancing the print trade. Sodium figured into the creation of caustic soda, used in soaps and an assortment of other goods, further "civilizing" the world by making it cleaner. Electrified machines and motors made industrial fabrication more rapid and efficient, leading to the improved production of iron, brass, copper, glass, porcelain, rubber products, asbestos, and mica. The increase in the number and variety of goods had a subsequent impact on Mexico as Americans and Europeans increasingly sought to sell their products abroad.[26]

Many of these goods came to Mexico through the port of Veracruz; as a result, the city made a fitting entry point for radio equipment. In addition to being the main hub for foreign trade, it had become a focal point of Porfirian progress and modernity. After years of off-and-on again progress, the Díaz administration hired the English firm of Sir

Weetman D. Pearson in 1895 to upgrade the ports. According to a number of observers, Pearson's work made Veracruz one of the finest ports in the Americas.[27] Steamships slowly traversed the murky waters. Sweaty-browed dock workers unloaded foreign products from boats, placing the goods on trains bound for the capital. SCOP employees had also been at work in the port. These federal agents had devoted time toward improving trade and communications facilities. Radio was their newest tool.

With the new wireless machines now in hand, Mena fixed his sights on lofty goals. Above all, he, like the younger Gutiérrez, imagined connecting the growing ports of the country and directing ships in the Gulf; he envisioned increasing Mexico's power. The general had kept abreast with radio developments in Europe and after much difficulty had imported equipment from the French inventor Edouard Ducretet, a renowned wireless pioneer. Operating out of his Paris laboratory, "equipped with a mast seventy-five feet high above the roof," Ducretet had started his commercial endeavors in 1898.[28] Based on the designs of another technological trailblazer, Russian scientist Alexander S. Popov, Ducretet's newest contraptions were already in demand.

His radios, however, failed to live up to the general's expectations. Experiments with the instruments started when Mena and González ordered the formation of a technical group, which began trials along the coast near the city of Veracruz in 1900. With frequent shipping traffic, the port was a logical place to set up the first operations. The devices worked satisfactorily at a distance of about one thousand feet. The transmissions, however, became more difficult to comprehend when further separated, and they failed altogether past four miles. These results shattered Mena's dreams of communicating with places like Campeche from Veracruz, or for that matter, even with ships far at sea.[29]

Mena contended he had the solution: he needed better equipment. Most specialists favored radios from the Marconi company of England. Mena explained to Congress that it had been a mistake to buy from Ducretet. Marconi's appliances were proven to be more reliable and powerful. Mena continued that the English apparatuses could transmit twelve and one half miles over land and almost ninety-five miles over sea. But as other SCOP employees reported, Marconi's equipment could not actually send messages at these distances, at least not reliably.[30] Still, many officials agreed that the English products would considerably best the six kilometers (3.7 miles) of Ducretet's machines. These hopes too ended in failure. Talks between state leaders and Marconi continued to break down over cost and service disputes.[31]

Consequently, the undauntable Mena focused his attention on inventors in Mexico. This time he approached Víctor Sauvade, a lesser-known inventor and French national who had lived in Mexico for some time. Not only did Sauvade provide new radio equipment—which he invented himself—he also came along to assist in the operation of the devices and to educate telegraphers, especially Alejandro Gutiérrez, on how to use them. SCOP employees additionally continued to work with the Ducretet products, including a Popov-Ducretet radiotelephone they had acquired. This latter group transmitted messages regularly along the coast of Veracruz between Hornos and the Isla de Sacrificios, a distance of two and one half miles.[32]

Mena was impressed with the Sauvade equipment. SCOP officials conducted a number of experiments in 1901 and 1902 with these radios in the port of Veracruz and between Boca del Rio and the Isla de Sacrificios (5.6 miles). They continued with trials between Boca del Rio and Isla de Enmedio (12.5 miles), Santiaguillo and Boca del Rio (18.5 miles), and the Fortaleza de Ulúa and Santiaguillo (23 miles), all with consistent clarity. They also made successful transmissions from the Fortaleza de Ulúa to the coast guard vessel *Donato Guerra*, which was moving at over twelve miles per hour at a distance of seventy miles, beginning the use of radio for communication with government ships in Mexico.[33] Mexican officials, like their Italian, British, German, Russian, and American counterparts, hoped to modernize communications in their navy, even if Mexico's flotillas paled in size and strength.

President Díaz's personal offices additionally became involved in radio. Toward the end of 1900, Reyes, on behalf of President Díaz, asked

FIG. 3. Roberto Jofre radio experiment, 1900. Courtesy of Acervos Históricos de la Universidad Iberoamericana.

Treasurer José Yves Limantour to release approximately forty pesos to buy parts necessary to build another, apparently cheap, experimental radiotelegraphy machine.[34] Maybe the small device was not expensive to build. Or, most of the device had already been constructed with other funds. Under the charge of the eminent medical doctor Roberto Jofre, radio tests began between Díaz's residency in Chapultepec and the National Palace.[35] It was in that December that Jofre and a crew of assistants sent the first documented radio message in Mexico. In honor of the president, it stated a simple and historically relevant message: "Congratulations on your sixth reelection."[36] Using another modern technology, photography, they took a picture of one of the devices and the group who made the experiment a success. Nothing else is currently known about what happened to this equipment.

Although reliant on foreign technologies, members of the Porfirian government conducted these operations: ambassadors abroad reported on the devices; military and SCOP officials obtained the first radiotelegraph machines and began the initial experiments alongside a French-immigrant inventor; and the president himself hired experimenters to build a radio device for the presidential palace. These actions show a government trying to reduce the influence of outside governments, contrary to resilient beliefs that the Díaz administration kowtowed to foreign powers.

Internalizing Empire

As the Chapultepec experiment indicates, corresponding with sea vessels and between ports were not the only designs for wireless communications. More than anything else, SCOP and military personnel used this technology in attempts to connect and control Mexico's frontiers. They wanted to combat previous failures to bring fringe regions within Mexico's political boundaries under the control of the central government. This lack of control presented a number of problems for Díaz. For one, these territories had resources that lay underdeveloped or in foreign hands. These same areas possessed local populations who promoted independence or failed to recognize Mexican authority. And finally, the border regions in the north remained threatened by an expansive United States, whose population, economy, and ambition appeared insatiable.

In Mexico, as with most other countries, radio applications built on previous designs for cable communications. The first wire telegraphy operations took place in the 1850s, and plans to increase telegraph lines to the frontier territories were in the works shortly thereafter. Construction of telegraph lines expanded at a much greater rate under the Díaz administration. Telegraphy became an indispensable tool for large businesses, the government, and the military. By the late 1870s, cables used for Morse-code transmissions connected the Pacific and Atlantic coasts. State agents expanded telegraphic operations into the northern and southern frontiers to thwart, or at least better respond to, internal rebellions and filibusters from "expansionist gringos."[37]

However, the construction of telegraph lines in the nation's hinterlands proved problematic. Not only were they farther away from main centers of capital, resources, and the federal government, the environment made the construction and maintenance of telegraph lines costly and extremely difficult. In the case of the Baja California Peninsula, the Gulf of California separated the territory from mainland Mexico except for a narrow, remote, and inhospitable strip of the Sonoran Desert. This arid and sparsely populated environment hindered telegraph construction. An attempt to lay a submarine cable across the Gulf, for some undeclared reason, failed as well.[38] State officials believed radio would remedy the communications situation.

The Gulf of California project matured under the direction of engineer Leandro D. Fernández, who replaced Mena as the head of SCOP in 1902. In addition to being a politician, he was also a scientist. Before and after his stint at the SCOP, he enjoyed a distinguished academic career. As a cabinet member he worked hard at expanding Mexico's infrastructure and was an enthusiastic supporter of radio. The success of Gutiérrez's experimental transmissions in Veracruz convinced Fernández to move forward with the construction of stations to unite the less-developed provinces with the core of the nation.[39]

Fernández's plan was to create a reliable means of communication across the Gulf of California that separated the two districts of the Territory of Baja California from mainland Mexico. The distance across this body of water was far less than that between the major ports along the Gulf of Mexico, which made the Baja project more feasible. Although Baja California had historically possessed a small population—the 1900 census estimated the population of the entire peninsula at only 47,082 people—the region had increasingly become important to trade in the Pacific, especially with the United States, and as a center for mining.[40]

Economic development was a driving factor in the advancement of communications between the mainland and the peninsula. Gold, silver, and copper had enticed a number of Mexicans and foreigners into mineral extraction endeavors in the area during the last half of the 1800s. Some of these enterprises became major developers in the region. The mining company El Boleo, for example, possessed a small fleet of ships and put substantial resources into fostering growth in the peninsula. Its employees built the townsite of Santa Rosalía shortly after Frenchman C. A. la Forgue founded El Boleo in 1885. Backed with substantial capital from the Rothschild family in Europe, El Boleo became one of three largest foreign-financed mining companies in Mexico and a major economic power on the peninsula. It extracted hundreds of thousands of tons of copper each year out of the Southern District of the Territory of Baja California. It exported much of this metal but sold some domestically to be used in such items as wires for telegraph lines.[41]

El Boleo additionally played a large role in the construction of the first radio station in Baja California, which the company used to its financial advantage. This assistance was important. Unlike subsequent radio-building projects that proved difficult in the Territory of Quintana Roo, El Boleo provided local expertise, equipment, and resources. This development enticed SCOP leaders to first direct their attention to the Baja California Peninsula. El Boleo, despite being foreign owned, also cooperated closely with the Mexican state, countering U.S. businesses that preferred to orient themselves with their home country over Mexico City.

Powerful political reasons also existed for building the first radio stations in Baja California and Sonora. Internal revolts rocked the region in the 1850s. In the same decade, foreign filibusters, driven by successful U.S. land grabs of Mexican territory during the Texas Revolution and the subsequent annexation of Texas into the United States (1835–36, 1845), the Bear Flag Revolt in Alta California (1846), and the Mexican-American War (1846–48), invaded both Sonora and Baja California in the 1850s. Most famous of these adventurers was William Walker, who after failing to establish the "Republic of Lower California," took power in Nicaragua in 1856. Similar plots cropped up in the 1880s and 1890s and continued into the twentieth century.[42] As late as 1908, *the New York Herald* was calling for the annexation of Baja California.[43]

Filibustering after the Mexican-American War consistently failed, but earlier policies of the Díaz's government and previous administrations had allowed northern Baja California and Sonora to become well within the economic orbit of the United States vis-à-vis capital, railroads, and telegraphs. Díaz dangerously planned to gain from this infusion of American money and technology while providing increased training for Mexican specialists. But even he had grown concerned. To balance the swelling presence of foreigners, he created stronger communications links between central Mexico and the northwest, an arrangement he hoped would help hold these more autonomous regions and eventually bring them under stronger federal control. Newspapers specifically cited the radio stations as essential to increased government vigilance over the peninsula to protect it from El Norte, promoting the construction of additional wireless and naval outposts.[44]

Under the direction of Fernández, González, and Enrique Schöndube—AEG's main radio representative in Mexico—work on the Gulf of California radio stations began in 1902. Schöndube had previously worked with the Porfirian government on a number of projects involving electricity, and he knew Díaz personally. Along with two other German engineers, Schöndube surveyed the area for the best place to build the stations. The original idea was to construct one radio outpost on Vigía Hill just north of the town of Guaymas, Sonora, and the second directly across the Gulf in the coastal town of Santa Rosalía. But following the German engineers' initial survey they abandoned the Vigía site after concluding that the mountainous terrain would interfere with radio signals. Instead, they selected a high point near the lighthouse at Cabo Haro. The Santa Rosalía site proved to be a good original choice, and construction began at both places in December 1902. The projects did not take long to complete. El Boleo and government employees finished the Baja California station on January 31, 1903. The project on the other side of the sea was completed on February 6.[45] After hiring telegraphers Luis Sánchez, Raymundo Sardaneta, Juan José Flores Treviño, and Pedro N. Cota to study radio and to operate the stations, the two posts exchanged their first *marconigramas* on February 16.[46]

Communications officials thereafter ordered the construction of wire telegraphs to connect the Sonoran radio station to central Mexico, which workers completed over the following year.[47] The radio stations ultimately accelerated a wave of construction of all available means of modern communication in the area.[48]

Unlike previous experiments, these Gulf of California stations were open to the public. The government wanted peninsular businesses and residents to communicate with mainland Mexico. The government charged a small fee for the service. In 1906 the cost to use the radio stations was the same as the other telegraphic operations that were under way in western Mexico: "One peso for the first ten words and ten centavos for each additional word."[49] The greatest users were most assuredly people associated with El Boleo, who dominated the Santa Rosalía area.

To the great excitement of Fernández, the stations worked well at first, though problems arose shortly thereafter. Radio equipment at the turn of the twentieth century was not consistently reliable. Most Mexican telegraphers had little real experience with the technology and largely learned on the job. Fernández admitted to Congress that the government's communications officials lacked mastery of the "theories and practices" of radio.[50] The first years of experiments, however, had taught officials a few things, mainly that atmospheric interference, despite technological improvements, occurred frequently from July to September when the rain and heat were most intense.[51] If green at first, radiotelegraphers gained critical technical expertise in these early operations. They would go on to become prominent in Mexican communications development during the Revolution and beyond.

Although problems persisted, the stations worked well enough to entice Fernández to go ahead with the construction of further works. In 1906, a pivotal year in terms of radio development, the same communications officials and German specialists continued to expand operations in the areas around the Gulf of California. This time they aimed at connecting the port of Mazatlán, Sinaloa—interestingly enough, home to an increasingly powerful German community—to the telegraphic offices in the Southern District of Baja California.[52] The engineers decided to erect the outposts at Cerritos, north of Mazatlán and next to the town of San José del Cabo. Once again, overland cables connected both stations to other telegraph offices in their respective states and into the federal system. These outposts took two years to construct.[53]

In 1905, the year before the Mazatlán–San José del Cabo project got under way, Fernandez's offices started the construction of radio stations in another provincial territory: Quintana Roo. For these operations they chose the small and newly established town of Payo Obispo (present-day Chetumal) and the also recently constructed naval port of Xcalak. Like the previous stations, SCOP officials used German equipment, though now from AEG's successor company—Telefunken.[54] Along with the naval facilities, the stations stood out in their rural tropical environment. The Quintana Roo stations possessed antennas "consisting of 36 lines in the form of umbrellas suspended from a tower of iron twenty-seven meters tall."[55] Marconi's radio towers had startled the natives of the Isle of Wight; surely the Maya and the British colonial woodcutters of the Rio Hondo river basin, further removed from industrialized centers, were even more surprised.[56]

Separated by the Bay of Chetumal, the two stations lay only thirty-seven miles apart, a reality making their construction a very workable proposition. Multiple other reasons existed for building these stations. As in the Territory of Baja California, residents in Quintana Roo had evaded central government control. The region was little developed and government officials saw radio as a convenient way to start incorporating the southern frontier. But these stations, unlike those in Baja California, possessed a strong military element. Not only were there continuing conflicts between the government and various Mayan groups, especially the Santa Cruz, but also increasing military conflicts in neighboring Central American countries.

In the late 1800s, many Maya communities in southwest Yucatán remained independent from the Mexican government. Their independence was a legacy of the Spanish Empire's failure to suppress and to incorporate the area. The early Mexican republic and the short-lived independent state of Yucatán (1841–48) fared no better. Political, economic, and cultural conflicts exploded a long and often brutal war in the region fought predominately between Mayan peasants on one hand and the Yucatecan and Mexican governments on the other. It remained ongoing from 1847 until the 1910s. Scholars commonly refer to this period, especially during the 1840s and 1850s, as the Caste War. The pacification of the Maya along Mexico's southern frontier remained a goal throughout Díaz's time in office. The final conquest of Chan Santa Cruz in 1901 by Porfirian general Ignacio Bravo and collaborating local leaders is sometimes regarded as the definitive end to the Caste War, however, segments of the region's population contested Mexican control well into the Revolution.[57]

Another important factor in establishing firmer control in the region was Mexican relations with the British Empire, specifically the colony of British Honduras.[58] For most of the 1800s, the British colonists reluctantly supported the Maya population in the area that would become Quintana Roo. The colonists cut valuable woods on Mayan lands in trade for weapons and other goods. As trade with Mexico became more important to Great Britain during the 1890s, however, a shift in policy took place. Officials in London informed the colonial leaders of British Honduras that they needed to comply with Mexican attempts to pacify the region. Between 1893 and 1897, Mexico and England formally agreed that the Hondo River was the boundary between the two countries. And in 1898 Mexican officials and businessmen began work founding a city along the Bay of Chetumal, Payo Obispo, and a naval site, Xcalak. The bay, in turn, became a main military base of operations, complementing Peto, which rested in inland Yucatán east of Chan Santa Cruz.[59] The towns became another motivation for British Honduran cooperation because they became an important source of trade.

Engineers and local workers built the Quintana Roo wireless stations at the two newly established settlements on the Bay of Chetumal. There were various reasons for selecting these locations. For one, Mexico had ordered its small southern flotilla to observe the bay and the Hondo River, which drained into it, with the mission of stopping British contraband to Maya rebels, especially weapons. However, Great Britain and Mexico had come to a disagreement over the allowance of Mexican ships traveling to Payo Obispo. The only passable channel through the shallow bay zigzagged across the international border and British officials feared that any decision allowing free passage to Mexican vessels would someday haunt them if Mexico continued to become a more powerful nation.[60] The radio station connecting the naval base and Payo Obispo at the other end helped alleviate some of the communications problems. It allowed Xcalak to send messages about naval operations and other matters to Payo Obispo, which could then be relayed to Peto—and in turn Mexico City—or to British Honduras and vice versa.[61] Military officials may have also hoped the stations could be used to communicate with portable radios carried by military detachments. The army had started experimenting with such portable devices beginning in 1906.[62] However, it appears—at least on record—that they did not use the devices in military campaigns until the first year of the Revolution in 1910 and 1911.[63] These Quintana Roo installations also complemented the incorporation of radio devices aboard Mexican coast guard vessels and the anticipation of continued advancements in naval communications.[64]

The construction of the Quintana Roo stations tested Porfirian can-do to the limit. For one, the appropriate materials were hard to obtain. There was no equivalent to El Boleo there. Communications officials often had to travel to Mérida in the neighboring state of Yucatán to acquire the essential products. It was also difficult to obtain enough workers and, according to Fernández, especially competent workers. Roads were rough, bugs were abundant, and the weather unpredictable. Builders, nevertheless, completed the stations by 1907.[65]

The radio stations in the frontier peninsulas of Mexico clearly exhibit Porfirian intentions for the technology. Radio allowed the Díaz administration to work on multiple coinciding goals: the centralization of state power, increased trade, building a stronger national identity, countering U.S. influence, and conquering autonomous regions within Mexico. Essential to these goals, the frontier stations allowed for the first rapid means of communication to these areas all together. There was much less motivation to build stations inland outside of the small experimental unit in Chapultepec. There was little immediate need. Telegraph stations in Sinaloa, Sonora, and Quintana Roo transferred radio messages to the cable system, which allowed for a rapid link to Mexico City and other urban areas. SCOP technicians were planning to build a high-power station in Campeche (it had been an aspiration since the 1902 experiments in Veracruz), but they would not complete such a project until 1910.

Beginning in 1906, private enterprises also started building radio operations, sometimes undermining state control. These transmitters and receivers were most common among American mining industries in northern Mexico. Motivated by the success of the government stations, George H. Holmes of the La Esmeralda mine obtained permission from the Díaz administration to start transmissions between his mine and the communities of Guadalupe y Calvo and Hidalgo del Parral, Chihuahua.[66] Most American mining businesses, however, were more concerned about communicating with corporate offices in the United States. Corporate communications certainly were the priority of stations constructed by the American Smelting and Refining Company (ASARCO), the Cananea Consolidated Copper Company, the Lluvia de Oro Gold Mining Company, and the owners of the Chispas mine, which operated illegally without government concessions. Belligerents during the Revolution would later close these radio operations because U.S. business interests used them to spy on combatants.[67]

The use of private radio spread beyond mining operations. A handful of businesses and hacendados obtained radio equipment to "communicate with neighboring populations" because storms commonly destroyed telegraph lines.[68] In response to individual and business requests, the Díaz administration created a specific department within SCOP to grant radio permits to private businesses. Officials specialized in communications and public works legislation drew up contracts with private entrepreneurs and experimenters, an arrangement allowing for some supervised private radio use.[69] The government, however, controlled the more influential public stations and, outside of Telefunken, remained resistive to foreign wireless companies.

While the SCOP was building stations in the territories of Quintana Roo and Baja California, the Porfirian government also participated in the 1906 International Radio Telegraph Convention in Berlin. The Díaz administration sent General José M. Pérez to represent its interests during the debates over global radio policies and Marconi's attempt to monopolize services. It was the second international conference on wireless communications, and it had expanded significantly since the first meeting in Berlin in 1903. The earlier convention only included representatives from France, Great Britain, Italy, Austria-Hungary, Russia, Spain, and the United States.[70] In 1906 there were delegates from twenty additional countries, including five from Latin America: Argentina, Brazil, Chile, Mexico, and Uruguay. In addition to working out an alphabet for international signaling, the meeting discussed matters of war, trade, and especially transmissions between ships and from sea vessels to shore. Precedents for the discussion were not only the first radio convention, but also the October 9, 1874, Treaty of Berne and the June 15, 1897, Washington Universal Postal Convention; both of which had focused on unifying disjointed international mail services and regulations.[71] These precedents indicate how many of the attending officials viewed radio services as an extension of public postal and telegraphic operations. However, there was a military element to Mexico's position as well. The general had specific orders to protect Mexican interests, especially in consideration of radio's relationship with coastal stations and ships of war.[72] Following the wishes of his superiors, Pérez sided against England, as most other representatives did, and with Mexico's German radio providers. This alliance with Germany would continue to influence radio development in Mexico until the 1920s.

Backing the agreement, the Mexican Senate ratified the terms in 1907, which the *Diario Oficial* published in Spanish and French in February 1909. This conference, along with the Constitution of 1857— which had established that the government could monopolize postal services—provided the ground rules for communications officials.[73] The telegraphers and engineers involved in the newly constructed Cerritos and San José del Cabo stations quickly switched their operating frequency in order to fit within the new international guidelines.[74]

After completing the Cerritos and San José del Cabo projects, state officials returned to the mission of expanding coastal stations along the Gulf of Mexico. The project became more viable with improvements in transmission distances. The DGTN particularly focused on the Yucatán peninsula in hopes of connecting the region to Veracruz and other gulf ports and of further incorporating southeastern Mexico into the federal communications system. These experiments carried huge trade implications, but they, like the other radio projects, also had ramifications for military and political consolidation plans.[75]

Stations along the Gulf of Mexico, however, did not achieve the desired success until the last year of the Díaz administration. After struggling to find the appropriate location to build the radio stations and expropriating land to build the towers, regular *marconigrafía* services commenced between Veracruz and Campeche in October 1910. Possessing newer equipment with finer receivers and higher-powered transmitters, these stations could transmit at a distance of five hundred kilometers (or almost 310 miles). Gutiérrez's dream had finally been realized. SCOP engineers exuberantly stated that their achievement made Mexican ports of "the first order."[76]

In 1910, the DGTN and the staff at the penal colony on the Isla María Madre off the Pacific coast of the state of Nayarit completed a station that communicated with San José and Cerritos. Not only did this radio

FIG. 4. Mexican radiotelegraph stations, 1910. Map created by author.

allow communications between the prison and the mainland, but it also provided news about ships in the Pacific and the Gulf of California.[77] It would become a military target during the Revolution. All in all, the Díaz government built ten stations.

Life for federal telegraphers at these early radio stations must have been fairly simple, if not boring at times. Although interaction with foreign traders and important officials must have been interesting on more than one occasion, these outposts were often in desolate locations. Employees commonly opened their doors to business at eight in the morning and generally closed around one in the afternoon.[78] Although often separate from nearby villages, these stations sometimes included additional facilities such as a warehouse, a small naval base, an oven for baking bread, a post office, a wharf, and a rain-catching system for drinking water.[79] Typical for the building housing the radio equipment was "a house of Iron and wood and two iron towers forty-five meters tall upon elevated land."[80] The telegrapher's house, like many others, included a kitchen and some furniture. The visible difference was the abundance of bulky wires and machines necessary for the job.[81]

Apparently the work could be dangerous as well. In February 1910, the manager at the Cabo Haro station ignited the radio building while cleaning a gas generator. After the fire exploded nearby cans of gas, oil, and alcohol, the blaze ravished most of the structure within minutes. The fire also destroyed all of the radio equipment, thereby suspending communications with Santa Rosalía until the government constructed another station at Bacochibampo near Guaymas in June 1910.[82] Still, even if the work was sometimes tedious, telegraphers were important to a number of powerful businesses and government leaders. While usually not an entrée into high society, the job of radio telegrapher offered stability and respectability. Unbeknown to them at the time, their lives would become much more interesting and endangered during the Revolution. Their services became highly sought after as successful communications could spell success or defeat for the belligerent factions. For competing armies, radiotelegraphers became the most useful compatriots and the worst of enemies.

Radio and Military Modernization

As is so often the case in technological development in general, the military played a crucial role in the advancement of radio. Matters of war, state security, and insurrection colored the views of military commanders who doubled as political leaders. As exhibited by the first frontier stations, these military officials, like their SCOP counterparts, hoped radio would help them consolidate state control, damper threats of American filibustering, and to finally quell Mayan rebellion. Together, the army and the SCOP started experiments with portable field radios, hoping to incorporate wireless communications as a part of a more modern military organization. These portable devices, however, were still in an experimental stage when the Revolution erupted in 1910. Although radio became better known and increasingly used by the last year of the Porfiriato, most army and naval units still relied on older forms of technology.

A company of telegraphers already existed in the army's Department of Engineers. In time of war it was organized into divisions of "ten telegraphers, ten orderlies, and ten horses." Some of their main tasks included "the construction, reparation, and destruction of paths of transportation," "the installation, conservation, exploitation, and destruction of telegraph and telephone lines, military as well as public and private," and "the establishment of posts for telegraphic signals of any system and the arrangement of the corresponding codes." They operated as the vanguard and rearguard of army operations, gathering and transferring information, preparing means of transportation, and preparing points of communication. In times of peace, they studied theory and applications useful for military service. The Department of Engineers additionally possessed a technical group that also studied the "reception and transmission of telegraphic signals and dispatches." The Military College likewise taught courses in campaign communications. Among the professors and technicians in the armed forces, radio had become a new and important information technology topic. Engineering, telegraphy, and naval students were at least introduced to the subject though rarely practiced in it.[83]

Radio operations in the military, as in other sectors of Mexican society, accelerated in 1906. The radio stations under construction in Quintana Roo served the army and the navy.[84] SCOP employees, including radio specialist Luis Sánchez Martinez, who would continue to work with wireless military applications throughout the Revolution, also began a series of field radio experiments for the military.[85] The goal was to build movable equipment that could successfully transmit messages at a distance of thirty kilometers (or 18.6 miles) over flat terrain. Officials built two sets from parts imported Europe and fabricated in Mexico. The resulting products must have been an impressive spectacle. Each one possessed a bronze antenna thirty-five meters tall, a "ground antenna, of six similar poles and about forty meters long, which rest almost horizontal," and "a dynamo of 45 volts per amp mounted on a bicycle frame to power it [the radio] by means of the corresponding pedals."[86] With the addition of numerous other pieces of equipment to the already large contraptions, the devices may have been "portable," but they were not easily moved.

The first experiments with these *estaciones portátiles* began in February 1907 in the vicinity of Mexico City. SCOP employees began by transmitting at short distances. They set up one station on the campus of the School of Agriculture and the other in a small community called San Alvaro, a little over a mile away. During the following months they moved the equipment to a number of subsequent locations, testing it at different distances. They discovered that the machines failed to receive messages at about fifteen miles. The resulting SCOP report concluded that the portable radios worked well enough at distances of twelve and one half miles, but not at the eighteen and one half miles that the military had hoped for.[87]

Most soldiers had little experience with radiotelegraphy. Despite the few portable devices, the Quintana Roo stations, and experiments on sea vessels, radio was a rarity. Even in the areas of the military where radio had the most applicability—the Department of Engineers, the navy, expeditionary forces—older technologies were more prevalent. Specialists learned how to set up and destroy wire telegraphic and telephonic operations and how to use flag signaling. In addition to light signals, most sailors still relied on flag communications.

The Military College, however, did not ignore radio. It trained select specialists in wireless communication technologies. "Electricity instructor" Roberto Torres Ovando, whom the Díaz administration paid to study wireless communications in the United States, meticulously kept up with radio advancements. He helped train a small cadre of officers who, in December 1910, would attempt the first use of portable radios in combat in Mexico. Torres was excited about the possibilities of radio for the army. So too were the editors of *El Imparcial*, the not so impartial Porfirian newspaper, which printed stories about how the military would benefit from portable radios like the ones used in Germany, Austria, and Russia.[88]

Conclusion: Porfirian Radio

By the end of 1910, radio still played a secondary role in communications in Mexico, as it did in all countries using the technology. Wire telegraphs and telephones were more relied upon for infrastructural needs. But radio proved essential to increasing the country's connection with the outside world and in linking provincial territories—Baja California and Quintana Roo—to the mainland. Political, economic, and military motivations drove the development of this technology, although a small group of hacendados, engineers, and academics experimented with radio as well. American mining industries, disconnected from far away corporate offices, also turned to radio as a new means of communication.

However, it was the SCOP and the military, with the assistance of foreign engineers and businessmen, who did the most to build a viable wireless communications system. State employees managed the public stations. They also operated the army and naval radio operations. These officials were in the process of making portable radios more efficient when the Revolution started in late 1910. With the outbreak of the subsequent civil war, especially after 1913, increased warfare within Mexico shifted the emphasis of radio use to fit the tumultuous circumstances. Although the Revolution eventually wreaked havoc on the economy, combatants' desires for better communications increased the importation of wireless equipment. As discussed in chapters 2 and 3, the war would become a training ground for many of Mexico's future radio specialists.

2 Radio in Revolution

It's the first time our army has used this prodigious invention in a
military campaign.

A radiotelegrapher in the Mexican army, December 1910

The day after Christmas, 1910, a weary anonymous American mining engineer crossed from Mexico into El Paso, Texas. A mixture of
anxiety and curiosity filled the streets. A revolution had erupted in
Mexico. Just the month before, another man, Francisco I. Madero, the
reform-minded son of wealthy landowners from the northern Mexican
state of Coahuila, had called the Mexican nation to arms from El Paso
after unsuccessfully running against Díaz in a fraud-filled election.
Rebels attacked government positions in the northern mountains of
Chihuahua, and the seven-term president had sent forces north from
Mexico City. The engineer was not the only U.S. citizen the escalating
violence convinced to return to the United States.

Upon the engineer's arrival to El Paso, he told reporters that he had run
into the federal forces of Brigadier General Juan J. Navarro, an old, deco-
rated military companion of Díaz. The Mexican president had charged
Navarro with punishing the insurrectionary upstarts. His soldiers,
however, lacked motivation. According to the American engineer,
Navarro's three hundred "sorry-looking fighters" looked demoralized after suffering defeat at battles in Aldana and Malpaso, where

they left many of their dead comrades unburied. The American also provided a second bit of intriguing news: another advancing force—the Ninth Battalion—carried portable radio equipment. Having established a wireless tower in Chihuahua, they hoped to restore communications with Navarro and the front lines of battle. The time had come for the military to test radio in war.[1]

The Mexican Revolution started out as a political revolt, quickly became a full-scale social revolution, and ultimately spiraled into a drawn-out series of civil wars. It resulted in approximately 1.5 million deaths and displaced people.[2] Although the most destructive violence ended in 1920, subsequent rebellions would continue to rock Mexico until 1938. Meanwhile, the revolutionary victors moved forward with plans to secularize, industrialize, and centralize the nation with mixed success. Although the leaders of revolutionary governments incorporated more people, they reverted to authoritarian practices as they attempted to secure their own power and put down insurrections.

During the Revolution radio became an important component of military intelligence and the logistical infrastructure of rebel groups and those who controlled the state. Wire telegraphy still dominated electronic communications, but radio increasingly played an important role as the "ears" of war. In fact, warfare accelerated much of the technological revolution that had begun in the mid-nineteenth century as combatants incorporated tools to better facilitate transportation and communication during their campaigns. This rapid growth was definitely the case with radio. Not only did the number of devices increase, but so too did the number of radiotelegraphers. This increase, however, was almost solely among militants. Growth was hampered among civilian experimenters deterred by the violence. Civilians transmitting radio messages were often seen as possible spies, and they possessed equipment valuable to militants. For the commanders of armies, wireless messages proved crucial at times, changing the course of decisive battles. Because of the war, military considerations would continue to color how leaders thought about radio in the subsequent decades.

The Madero Rebellion and the Continuation of Porfirian Wireless Policies

The Madero uprising and subsequent presidency started as a political revolt in 1910 and ended with Madero's assassination in February 1913. Like Díaz himself, much of the Porfirian administration—cabinet members, advisers, military officers, governors—had grown old and stingy with their power. A new generation of middle-class professionals in league with ostracized elites variously agitated for a greater voice, for democracy, or at least political spoils. Few of these middle-sector discontents were genuinely radical; rather, they saw themselves as champions of the ideals of the 1857 Constitution, ideals to which Porfirian political leaders had shown only lip service.

It was in a 1908 interview with U.S. journalist James Creelman that Díaz inflamed political fires to a point where they could not be easily extinguished. Telling the reporter that he would not run in the 1910 election, Díaz's comments spurred a burst of widespread civil activity. The dictator's decision to betray his word antagonized his now energized opponents. By 1910 Madero had solidified his position as the candidate of the Anti-reelecionistas, a newly formed opposition party. He had gathered enough support to prompt Díaz to order his arrest before the presidential contest. After escaping incarceration, Madero proclaimed his rebellion while in exile in Texas, though he stated that it was written in the Mexican city of San Luis Potosí to avoid problems associated with inciting revolt from abroad. His uprising fared poorly at first as most of his urban intellectual followers met early defeat. That might have spelled the end of his revolt if a large number of disgruntled rural Mexicans had not also responded to his call to arms. Leaders from the northern mountains, including Pascual Orozco, José de la

Luz Blanco, and Pancho Villa gained armies of peasants, laborers, and small ranchers who battled the federals to a standstill, breathing air into the struggling revolution.[3]

It was under these conditions that Díaz sent Navarro's forces north. They made their way during the cold of winter to fight with rebels in the arid mountains around the city of Chihuahua. A number of witnesses made note of their appearance, including the U.S. mining engineer. After some early federal victories, insurrectionist forces inflicted an obvious and serious defeat to federal forces. Not only had these battles, which occurred on December 17–18, 1910, killed dozens of government soldiers and demoralized many of those that survived, the attackers had succeeded in isolating the federals, disrupting the wire telegraphs and the nearest railway line. In response, Díaz demanded the accounts of every surviving officer involved and sent the Ninth Battalion to Chihuahua.[4]

The defeat prompted Díaz to test one of Marconi's earlier sales pitches for radiotelegraphy, using the technology to connect with troops in a "lonely spot" where rebels had cut telegraph lines.[5] General Angel García Peña, head of the Geographic Exploration Commission, tagged along with the Ninth Battalion, bringing radio materials to "establish and experiment with TSH [wireless] in order to connect Navarro to this city [Chihuahua]" and other "various points in the state."[6] Díaz had already planned on incorporating radio devices into the northern campaign, but the losses created a new sense of urgency.

In the three-year period (1907–10) of experimenting with field radios, the army and SCOP had made a number of improvements. They had reached greater distances, reportedly over eighty miles, and they had built special carts specifically for carrying the machines. Military telegraphers were enthusiastic about testing the devices in real war conditions. But radios were still far from reliable, especially in the mountainous terrain of northern Mexico. Operating within the forces of General Gonzalo Luque in early January, García Peña could not get the radios to work. He was not sure why. According to him, it was either atmospheric conditions or insufficient power. Despite the failure, he planned to get the radios into a better operational state. The rapid deterioration of the Díaz government cut his plans short.[7]

Rebels, too, understood that radios, when they worked, could provide important benefits, though access to equipment was extremely limited. In addition to the northern rebels, Emiliano Zapata led an uprising just south of Mexico City in the small central state of Morelos over the loss of peasant lands. Hailing from the small town of Anenecuilco, Zapata came from a family of community leaders. He would become a tenacious and sincere revolutionary dedicated to the people and land of Morelos.[8] Nominally joining forces with Madero, Zapatistas took advantage of radio via an American reporter who had access to a transmitter. After they captured Cuernavaca, the state capital of Morelos, Stephen Bonsal, a popular Latin American and Caribbean correspondent for the *New York Times*, stated on May 22, 1911, that "Cuernavaca is quite orderly under the rebels, and foreigners are reassured. As an advance on wireless telegraphy I would state that the revolutionary authorities in the city reported to me the capture of Cuernavaca five hours before they took the place, but they made good on their word."[9] It is unclear what device Bonsal used. It is likely that he relied on a government or U.S. embassy station. However Bonsal sent his messages, rebel leaders in Morelos, contrary to popular notions—then and now—understood that foreign reporters and radio relayed information and influenced foreign public opinion.

Back in northern Mexico, the rebels under Madero won a hard-fought victory at Ciudad Juárez along the Mexico-U.S. border. Against Madero's orders, his forces attacked the city, ultimately taking it. This rebel victory, in addition to that at Cuernavaca, led to the downfall of Díaz and, subsequently, to years of factional fighting.[10]

With the fall of Ciudad Juárez and Cuernavaca, Díaz read the writing on the wall and resigned. Uprisings had sprouted up across the country. He boarded the steamship *Ipiranga* and set off for exile in Europe. Within days of Madero's ascension an earthquake hit central Mexico, an uneasy sign for the new government. Díaz received news of the disaster via an

Atlantic City radio station. He gave his laments as he sailed away—both over the waves.[11]

Rebellion, however, did not stop with Madero's presidency (1911–13). Old revolts continued and new ones arose as Madero failed to enact the social changes that so many of those who fought under his banner demanded. The reluctance to make quick or extreme decisions, especially in matters of land distribution and municipal autonomy, provoked the Zapatistas to continue their fight against the federal government. Former Maderista general Pascual Orozco rebelled in northern Mexico in 1912 with the support of many of his original followers, but also from powerful Porfirians including Luis Terrazas and Enrique Creel. The rebellion made for an interesting mix of reactionary forces with revolutionaries who sought greater reform. Counterrevolutionaries also took up arms under the banners of Félix Díaz, the deposed dictator's nephew, and Bernardo Reyes, the former Porfirian secretary of war and governor of Nuevo León.

Madero understood the potential of radio, and he used it to counter the continuing rebellions. Commodore Manuel Azuerta reported via radio the outbreak of Felix Díaz's counterrevolutionary effort in Veracruz. Using equipment on an American Ward Line ship in the harbor, Azuerta contacted the Madero government in Mexico City. Via the same radio, he obtained orders from the capital. Along with the responses of Generals Joaquín Beltrán and José Hernández, Azuerta's actions were crucial to the quick capture of Felix Díaz, who was sent to prison in Mexico City alongside Reyes, whose previous revolt had also met defeat.[12]

The Madero administration installed wireless devices aboard its own ships as well. In 1911, communications specialists finished installing a radio on the gunboat *Bravo*. There had been plans to equip the ship, among others, since before the Revolution.[13] The *Bravo* was one of the Mexican navy's best ships. As one of the navy's vessels that voyaged to other countries, wireless apparatuses made communication with foreign nations and ships easier. The next year technicians installed a similar device on the *Melchor Ocampo*, a coast guard vessel that monitored Mexican shores along the Gulf of California.[14] Officials hoped that these boats would relay important intelligence to coastal stations and to the capital.

On land, the army under Madero—mostly the same forces that served Díaz—installed a French-made transmitter and receptor in the forest surrounding Chapultepec Castle in Mexico City and in the city of Torreón, Coahuila, exclusively for the use of the armed forces.[15] The Chapultepec station, overseen by engineer Estanislao González Salas, director of works of the National Palace and Chapultepec, consisted of "two elevated towers and a little house in the center for the employed operators" located in the northern part of the forest near the boundary of the park.[16] In addition to field radios, these stations bolstered communications between the president and the front of (counter) revolution rekindled in the north by Orozco. Through these stations, Madero and the military improved upon one of the communication deficiencies of the Díaz administration whose Chapultepec equipment was used solely in low-power experiments. Although the federal army under Madero had difficulty in decisively defeating Orozco, these radio posts helped the government maintain control in the Torreón vicinity by maintaining a constant communication link between it and Mexico City.

Outside of Torreón, the wireless offices surrounding Baja California were the only government stations close to continued upheaval. Taking advantage of the momentum provided by Madero, members of the radical *Partido Liberal Mexicano* (PLM), or Mexican Liberal Party—nominally led by intellectual Ricardo Flores Magón from his armchair in Los Angeles—in addition to members of the Industrial Workers of the World and foreign soldiers of fortune, ignited a rebellion in Baja California with aspirations for adventure and the creation of an anarchist utopia. Although the local rebels and invaders from just over the border, who saw themselves as liberators, successfully captured the cities of Algodones, Tecate, and Tijuana, the different factions and motives within the rebel force turned the Floresmagonista invasion into a farce that ultimately caused residents of Baja

California and leftist groups in the United States to abandon support for the movement.[17]

When former Porfirian diplomat Francisco León de la Barra became interim president (1911)—Madero had waited for formal elections to hoist him into high office—Flores Magón, like Zapata and Orozco, rebelled against his tentative former ally. News about continued Floresmagonista actions reached Mexico City via telegraph connections through the United States, but also via the Santa Rosalía and Bocochi-bampo wireless stations. The head of the Santa Rosalía *rurales*, a rural police force first created in the late 1860s, used the local station to report that approximately "sixty filibusters were marauding around Santa Rosalía and Camilla," an area that Flores Magón more or less correctly argued was controlled by El Boleo, "a rich French company."[18] Radio-telegraphers also noted the arrival of the steamship *Korrigan*, from which federal forces disembarked to Juventino Rosas's famous waltz "*Sobre las Olas*," or "Over the Waves."[19] The federal forces defeated the remnants of the PLM soldiers in the summer of 1911, thus ending the invasion.

Despite improvements, the Madero rebellion ultimately brought little change to the course of radio development. The advancements in military communications were already in planning before Madero took office. Employees and engineers contracted by the DGTN continued with the projects that they had already begun. Upgrades to the new SCOP building continued, receiving praise from Madero and his Secretary of Communications and Public Works Manuel Bonilla.[20] The lack of significant change in radio policy provides another example of the limited nature of Madero's "revolution."[21] It also exhibits the importance of the cadre of wireless specialists trained during the Porfiriato. They were the only people capable of carrying on the task of building Mexico's wireless infrastructure, and there were few people within Mexico who could replace them.

The biggest changes came in top leadership positions. Although new, Bonilla's job required him to quickly learn national and international communications policies. Soon after obtaining the position, he proved to be a capable negotiator during a dispute with the New York and Cuba Mail Steamship Company of the United States and the U.S. ambassador to Mexico Henry Lane Wilson, Madero's implacable foe. On October 1911, one of the American steamers had entered the waters around Veracruz. Communicating to the newly functional wireless station, the ship's telegrapher discovered that their calls were unwelcome. Miguel A. Cosio, then the managing officer at the Veracruz radio office, responded that under orders from a 1909 circular, Mexican stations were not to communicate with vessels from countries that had not ratified the provisions of the 1906 International Radio Telegraph Convention of Berlin except in cases of emergency. Cosio then recommended that agents of the American steamship company contact the U.S. government in order to work out an arrangement. Infuriated, Assistant General Manager W. D. Macy wrote a letter to Ambassador Wilson, hoping to resolve the issue with the Mexican government. Macy argued, and Wilson reiterated, that even though the United States was not in compliance with the 1906 International Radio Telegraph Convention of Berlin, Mexican officials should provide the New York and Cuba Steamship Company special communications privileges since it possessed a contract with the Mexican government to deliver mail abroad. Macy contended, however, that the real issue was that agents of Marconi's Wireless Telegraph and Signal Company had built the stations. According to his logic, Mexican stations were in direct competition with the United Wireless Company, a U.S. outfit that sold equipment to the New York and Cuba Steamship Company. Marconi's agents had convinced Mexican officials not to communicate with U.S. vessels.[22]

Bonilla responded firmly, displaying his dismay of Macy's ignorance. Of course, Macy's error in reasoning made it easier for Bonilla to refute the accusations. He argued blatantly that the Mexican government "never had accepted foreign influence" and that Mexican radio telegraphers held no obligation to Marconi.[23] Easiest to refute was the claim that Marconi had built the stations, which was just wrong. Telefunken had helped build them.

Mexican communications officials had also attempted to strictly apply the Berlin regulations, including in matters of adjusting wave frequencies of radio stations and the liquidation of account balances between participating countries. Since the U.S. Senate failed to ratify the provisions of the convention, Mexico had no legal means to ensure that American citizens pay the required tariffs asked by the Mexican government and stipulated in the 1906 conference. Still, Bonilla was willing to work with the U.S. government on the issue bilaterally if this fiscal component could be agreed upon. American ships provided an important means of international communication for Mexico. Bonilla proved to be a capable negotiator who failed to be intimidated by the arrogance of Macy and Wilson. Lucky for Bonilla, he also had a small but knowledgeable group of communications officials that had worked with radio technology over the previous decade.

Despite transitions in the leadership of SCOP, the work of the engineers and telegraphers below them remained the same. Most of the government stations rested in areas little touched by the rebellions in 1910 and 1911. As previously planned, experts including engineer José de Prida and foundational telegraphers including Juan José Flores Treviño and C. Alejandro Gutiérrez—with the assistance of German and Norwegian electronics specialists—put the finishing touches on the Veracruz and Campeche stations. De Prida additionally worked on improving the equipment at Payo Obispo, which by mid-1911 was in consistent contact with the Campeche operation, and in turn, Veracruz.[24] All were new accomplishments; none were new goals.

Even with the injection of war, SCOP officials continued to work on linking the frontiers with Mexico City and the rest of the nation. Replacing the incinerated Cabo Haro post, they opened the doors to the Bocochibampo wireless office in Sonora, once again establishing radio communications between the Baja Peninsula and the mainland. The DGTN was additionally working on establishing four other stations in San Quintín, Bahía Magdalena, Loreto, and Miramar—all on the Baja California Peninsula. As with the first radio towers, these operations were to be connected with wire telegraphs that linked the remainder of the territories.[25]

Federal radiotelegraphers reached other important benchmarks during the de la Barra and Madero administrations. In 1911 the DGTN put into action a plan to better organize the national radio system. For the first time, the government demanded systemized call letters for the country's wireless operations: XA for coastal stations XB for merchant ships, and XD for internal stations.[26] Perhaps most impressive, radio operators reached new distances with their transmissions from Veracruz and Campeche. With new German equipment in hand, they transmitted messages over five hundred miles during the day and over eight hundred miles at night (radio waves travel further at night because of better ionospheric conditions). These operations could reach the Isla Madre María, San José del Cabo, and Mazatlán posts. The gulf stations also allowed for international communications, establishing contact with Havana, Cuba; Pensacola, Key West, and New Orleans in the United States; and Colón, Panama.[27]

However, Madero and the new SCOP leadership cannot solely be credited with the accomplishments in wireless development. Almost all of these initiatives were part of plans specialists and government employees drew up during the Porfiriato. Most of these engineers and telegraphers remained loyal to the state—more so to their jobs—even though the top leadership changed frequently. Needless to mention, insufficient electronics and communications specialists existed to replace the DGTN workers who had built the Porfirian radio system, even if Maderistas had wanted to, which they did not. These experts and experimenters had made notable strides and their work remained solicited.

Nevertheless, federal employees did make impressive gains. Their work from May 1911 to February 1913 focused on organizing the national radio system, completing and improving Mexico's coastal stations, and equipping the navy. As during the Porfiriato, these projects were aimed at obtaining greater control over Mexico's fringes, improving trade, and projecting a modern appearance to the world. However, as exhibited by the military stations in Chapultepec and Chihuahua, the continuance of armed rebellion shifted some of the focus on radio development toward putting down the very uprisings that Madero himself had inflamed.

The Usurper and the Militarization of Radio

On Valentine's Day 1913, a radio report circulated from Mexico, to Cuba, to the United States claiming that someone had killed Madero. Upon obtaining the news, the editors of the *New York Times* printed an article on the Mexican president's death the same day.[28] The only problem with the account was that Madero was still very much alive. Madero was, however, murdered eight days later by a disgruntled army major following a ten-day long coup led by supporters of Félix Díaz and Bernardo Reyes, prominent members of the old Porfirian order. Following the release of Díaz and Reyes from Mexico City prisons by their followers, the overthrow attempt turned into a battle in the capital's streets. Madero was ultimately betrayed by Victoriano Huerta, the general whom the president had entrusted to defend the government. After working out an agreement with Díaz and American ambassador Henry Lane Wilson—Reyes had accidentally got himself shot on the first day of the uprising—Huerta assumed the presidency on February 20. The following night, a small group of soldiers executed Madero and his vice president, José María Pino Suárez, just past midnight, outside the Lecumberri Prison.

The premature report of Madero's death highlights one of the problems associated with radio, or most any other rapid means of communication for that matter: people are often wrong. A number of subsequent stories quickly revealed the error of Madero's demise, but the mix-up demonstrates that radio often circulated unconfirmed rumors; faster news did not mean correct news. Radio increased the number of reports coming from U.S. journalists in Mexico and other parts of Latin America, but they often confused as much as clarified events for U.S. policy makers and newspaper readers. Accuracy of

FIG. 5. Victoriano Huerta and his cabinet, c. 1913. Courtesy of Library of Congress, Prints and Photographs Division, Reproduction No. LC-DIG-ggbain-14712.

reports remained a problem for intervening U.S. government officials throughout the Revolution.

In Mexico, with the ascension of General Huerta to the presidency, military matters dominated radio development, as with almost every other aspect of his government. Madero had already increased wireless operations for the suppression of rebellion, but communications became a key element of Huerta's plans to secure his position and to tame the spreading violence. With militarization in mind, he invested in these apparatuses at a quicker pace than his predecessors. He saw radio as an indispensable tool for winning the war against the men and women who took up Madero's banner. The new rebels had significant support. Conflicts with the U.S. government, including incidents involving the establishment of U.S. radio stations in Mexico, made Huerta's position all the more difficult.

Huerta spoke about radio advancements during his first address to Congress on April 1, 1913, only a month and a half after Madero's assassination. Interestingly, he stressed that service was under way between the wireless operations in Campeche and Veracruz and the New York and Cuba Mail Steamship Company. Although U.S. and Mexican officials worked out the previously mentioned issues surrounding this company when both nations agreed to the provisions of the 1912 International Radiotelegraph Convention in London, it is notable that the topic was one of first things about radio that Huerta mentioned.[29] This move surely appeased Ambassador Wilson and constitutes one of Huerta's initial attempts to warm to the United States.

The general, however, quickly ran into his own conflict with the Northern Colossus. The 1912 U.S. elections ushered in President Woodrow Wilson, who disdained Huerta's coup and the possibility of U.S. complicity. Only months after Wilson's election, a conflict erupted over the placement of wireless devices in the U.S. embassy at Tampico, further increasing tensions. This poorly planned event caused an anti-American uproar in early July 1913. U.S. Marines, in uniform, disembarked from the warship *South Carolina* to install the equipment, causing public unrest. This hampered Huerta's attempts to appease the United States. Although some historians have argued that anti-American protest in mid-1913 "represented officially-orchestrated reactions to crises in Mexican-American relations" (and surely this was the case in many instances), this Tampico incident was genuinely popular in nature even if the Huerta regime tried to use it afterward to rally domestic support.[30] People in the gulf ports, like many government officials, had become wary of the American navy, which had floated off the shore since 1912, a presence that had been brought about not only by the recent revolts but also by constant complaints by U.S. ambassador Henry Lane Wilson and William W. Canada, the U.S. consul in Veracruz.

The establishment of an American wireless station in Tampico was a nonissue turned into a loud mess by the negligent actions of the U.S. armed forces. In mid and late June 1913, Huerta's minister of foreign affairs, Francisco León de la Barra, the SCOP leadership, and the Wilson administration worked out a deal to allow the establishment of the radio office in the American consulate in Tampico so that it could communicate political, economic, and security interests with American warships, the consulate in Veracruz, and the embassy in Mexico City.[31] In early July, however, the disembarking of disallowed, uniformed marines caused a "very bad impression among the public."[32] It was only after this public protest that the governor of Tamaulipas and other federal officials withdrew the permission to install the radio equipment. Huerta, upset with the U.S. government for stalling recognition of his dictatorship and with America's increased military presence, may have capitalized on the anti-American outbreak, but he did not initiate it. His administration did order the removal of the devices, antagonizing American officials who were "loath to believe that . . . [Huerta's] government would place obstacles in the way of prompt and adequate protection of its [U.S.] citizens in the Consular District of Tampico."[33] A month and a half later the Huerta administration agreed to reconsider the matter as long as no member of the U.S. armed forces, in uniform or not, installed and operated it.[34] The decision proved to be one that Huerta would regret. This little-recognized event strained tensions and foreshadowed future encounters. Thereafter, public sentiment against the United States along the Gulf of Mexico ebbed and flowed but nevertheless persisted. It would reach new heights during another incident involving the U.S. navy in Tampico and the U.S. invasion of Veracruz in 1914.[35]

Huerta's frustration with the United States mounted, and he increasingly drummed up anti-American sentiment as Wilson's opinion of him became clear; meanwhile, U.S. operatives continued to use wireless communications—on land and ships—not only for American residents in Mexico but also for military and diplomatic matters. The Wilson administration fed on these transmissions, trying to sort out the divided opinions received from diplomatic and military personnel. The navy was especially crucial to Wilson's intelligence-gathering operations. Secretary of the Navy Josephus Daniels was close to the president, and wireless devices, like those on the battleship *New Hampshire*, transmitted important matters relating to the situation in Mexico, including Huerta's threat not to accept their esteemed passenger, President Wilson's new replacement for Henry Lane Wilson, John Lind, in August.[36] Indeed, throughout the Revolution, U.S. warships and consulates used radio to relay messages about Mexico to the United States, increasing communications and news of revolutionary events, but offering a wide array of contradictory perspectives.[37]

Actions from the U.S. Pacific Fleet exhibited these incongruities. The American navy had been "observing" the Revolution from the Gulf of California just as their fellow sailors did in the Gulf of Mexico. Interestingly, the crews in the Pacific showed less hostility to Huerta's forces than the flotilla in the Gulf of Mexico. In fact, Huerta's ships clearly benefited from the former's wireless messages. The Constitutionalist navy in the Gulf of California—consisting of one ship, the *Tampico*—possessed radio facilities. The ship and its crew switched to the Constitutionalists' side in February 1914. Fernando Sánchez Ayala, a radiotelegrapher who had worked for SCOP during the Porfiriato and for the Huerta administration during 1913, installed and operated the equipment.[38] The remaining federal gunboats in the Gulf of California, especially the *Guerrero*, moved quickly to punish their traitorous brethren. On March 31, the *Guerrero* attacked the *Tampico*, heavily damaging it. The *Tampico* limped back to the small Sinaloan port of Topolobampo, where sailors and local rebels mended the vessel. Carrancista general Álvaro Obregón toured the ship while it was under repair in mid-April 1914. He hoped to get the *Tampico* back in operation for an upcoming siege of Mazatlán.

The crew of the *Guerrero* had different plans for the *Tampico*. Shortly after Constitutionalist craftsmen repaired the *Tampico*, the *Guerrero* attacked it again on June 16 off the shore of the Isla de San Ignacio between Guaymas and Mazatlán. This time the *Tampico* sank.[39] The American Pacific Fleet had tipped off Huertista general Joaquín Téllez via a "wireless bulletin" that the rebel ship had taken back to the waters in order to aid the siege at Mazatlán.[40] Téllez then ordered the

Guerrero out to confirm the American message. The U.S. navy watched the subsequent encounter, idling nearby.[41]

American warships constantly observed Mexican naval movements. U.S. officers interacted with their Mexican counterparts and rescued American citizens.[42] Press agents on board relayed combat news to the naval radio station in San Diego, and in turn, the U.S. press. One American correspondent on the USS *California* reported that just moments before the sinking of the *Tampico* an American lieutenant from the American destroyer *Preble* had "inspected" the rebel ship. The rebel crew had asked for American assistance with a tow or ship repairs, but the U.S. sailor refused, saying that such an action would counter American neutrality. Afterward, the officer stated that he "found the vessel pitifully unprepared for a fight. The guns were without sights and there were only one hundred rounds of ammunition. The ship's boiler had been burned out by the inexperienced crew, and the commander of the ship was crippled as a result of an accidental bullet wound in the leg." But "for four hours the *Tampico* made a gallant struggle."[43] Shortly thereafter, the *Guerrero*'s radio operator contacted Rear Admiral Thomas B. Howard, commander of the U.S. Pacific Squadron, stating that his ship was going to move north to Guaymas. The cruiser *Denver* accompanied the federal vessel for much of its voyage, which the *New York Times* relayed as "evidence of the friendly relations existing between Mexican federal officials and the United States naval forces."[44] After intercepting and partially decoding American messages about the U.S. invasion of Veracruz, Huertista naval officials in the Gulf of California became "very confused" at the increasingly contradictory international situation.[45] Something did not make sense. President Wilson may have set his aims against Huerta, but American naval officers showed little ill will in the Pacific.

Despite the actions of the U.S. Pacific Fleet, Wilson not only saw Huerta as an illegal usurper of power, but also a militarizing and—contrary to the hopes of many conservatives in the United States and

Mexico—a destabilizing force. Huerta went against the long-term interests of the United States. The Wilson administration put in place an arms embargo against him and instead slowly fed his opposition.

Huerta, as a result, imported a record amount of war materials from Europe and Japan, including wireless devices from the former. Huerta hired a German company in January 1914 for the erection of a wireless station "to maintain communication with the columns of Federal troops operating against the rebels in various parts of the country . . . because ordinary telegraph wires are so frequently cut."[46] He additionally bought wireless apparatuses from France. As early as June 1913, as Americans were attempting to get the Tampico embassy station approved, Huerta was ordering ten radio stations with twenty horsepower engines and one hundred meter–tall towers from Harlé & Cie. of Paris. Harlé sent the equipment in mid-1914.[47] The secret nature of radio operations by the military preceding and during Huerta's rule makes it difficult to determine the precise number of radios in use, but Huerta definitely bought more equipment in a shorter period of time than his predecessors. European officials, for their part, agreed that Mexico needed a strongman, and in turn overwhelmingly supported Huerta's regime. Indeed, European military and communication equipment sales to Huerta deliberately—if only briefly—challenged American influence in Latin America and isolated the Wilson administration diplomatically.[48]

Technological domination had been a key feature of state power before and during the Revolution, and Huerta, even more than former presidents, expanded radio as a means of control. Huerta's use of radio, however, was aimed less at connecting frontier peninsulas to Mexico City and more toward defeating his military opponents. In response, Huerta's enemies gathered their own weaponry and built their own communications networks.

Radio Contra Huerta

The escalation of fighting after Madero's assassination led to a substantial increase in the importation of radio equipment. Not only did the Huerta regime accelerate its incorporation of radio, but so too did the Constitutionalists, the rebel forces under "First Chief" Venustiano Carranza who picked up the banner of the preceding Madero revolution. Unlike Huerta, Carrancista forces mostly obtained wireless materials from the United States, initiating the beginning of serious U.S. radio sales to Mexico, showing the permeability of the border, and revealing Wilson's ultimate decision to tentatively back Huerta's opposition. Germany initially supported Huerta. The infusion of U.S. radio equipment proved decisive in battles that directly led to Huerta's ouster from office.

In many ways, the Constitutionalists began where the Maderistas had left off, but their leadership built a much larger coalition of forces less willing to compromise with members of the old regime. They were also more uncompromising in their war against Huerta. Carranza had no desire to end up like Madero. The First Chief hailed from a wealthy family from the northern state of Coahuila. He had worked his way up the political ladder to the position of senator, with the blessing of President Díaz in 1904, and later became governor of Coahuila during the Madero presidency. Following his proclamation against Huerta, Carranza declared himself the rightful heir to the Revolution and the national government, thereafter working diligently to unify wide-scattered revolutionary forces of various allegiances. Initially limited on funding and supplies, he never completely succeeded in obtaining the loyalty of all fighters opposed to Huerta. He did, however, ultimately manage to organize three armies of significant power and organization, all hailing from northern Mexico—the Army of the Northeast, the Army of the Northwest, and the Division of the North.[49] These forces were led by Pablo González, Álvaro Obregón, and Pancho Villa, respectively.

Carranza moved quickly to incorporate radio to help organize his command. Within months of Carranza's call to arms, he and allied anti-Huerta insurgents possessed a radio station in Hermosillo, the capital of Sonora, where a wireless office had been functioning by late 1913. After capturing the city of Chihuahua on December 8, 1913, Villa oversaw the construction of a sister station. Carranza and Villa used the operations to spy on federal and American radio transmissions.

Radio technology had an impact on all three of the main Constitutionalist armies. González established a wireless office on his train of war. According to the memoir of his secretary Manuel W. González, the telegraph car adjoined the general's personal quarters. It was well kept and fitted "with bunk beds for the telegraphers and some of the officials of the general staff" and a telegraph station. His radio operator was well liked, tall, fat, and a bit of jokester. Both Pablo and Manuel González valued his work as highly as they did his likable personality.[50]

Of the Constitutionalist forces, the Army of the Northwest possessed the least amount of radio equipment. No record to date shows that Obregón used field radios during his slow and methodic operations along the Pacific coast. He never mentioned possessing this type of equipment in his memoirs. He did write about using telegraphy regularly, indeed that it was crucial to his operations. Some statements infer the possibility of wireless use. Obregón wrote that Wallace Buctoll of the U.S. Pacific Fleet had contacted him in late April 1914. In June, Obregón told Villa—then in a dispute with Carranza—that he was unclear on the exact details of their conflict because the continual rains had hindered the clarity of his telegraphic service.[51] These suggest, though they do not definitively prove, that Obregón used radio. Either way, he most certainly understood that radio was a powerful tool that could change the course of military outcomes.

Obregón's recognition of radio's power is clearest in his successful attempt to sabotage an enemy station. On June 4, 1914, Obregón reported to Constitutionalist General Rafael Buelna that he had dispatched Captain Medina Cruz of San Blas, Tepic, to overtake the small federal outpost in the Islas Marías with the goal of destroying its radio office. After commandeering the steamship *Union* that evening, Medina and a small force landed on Isla Magdalena—the island with the wireless post—early before sunrise the next morning. There, he and his small contingent of soldiers overtook the ten federal soldiers and their commanding officer. Immediately afterwards, Medina's men disabled the wireless outpost by sabotaging the connection between the station's motor and its electrical source. Obregón commented that by removing this means of communication from the federal arsenal, Medina had disabled an operation that greatly jeopardized their actions.[52]

Villa proved impressively keen in understanding communications and transportation networks. Historians have noted Villa's smart use of trains, telegraphy, and force during his capture of Juárez, Chihuahua, following his failed first attempt to take the state's capital. Although federal forces assumed he fled south to lick his wounds, he actually seized a train to the north of the city, forcing by gunpoint a train-station telegrapher to tell Juárez officials that the line to the south was destroyed. In turn, the Juárez operators told the train to return. Villa loaded up his men, violently convincing telegraphers along the way to give the clear.[53] He captured the city without a fight. Villa installed a radio station there the following year. Like General González, Villa constructed a mobile device on a military train for his personal use, and most importantly, he used radio during his decisive campaign into central Mexico in the spring of 1914.

Wireless communications played a determining role in one of the most important battles between Huertista and Villista forces and, for that matter, of the revolution—Torreón, March 1914. This battle directly led to Huerta's resignation from office. Both Huerta's generals and the rebel forces relied heavily on radio communication not only to give orders but also to make important tactical decisions. Transmissions overheard by Villa's wireless operators initiated the movement of forces at crucial moments. On March 15, Villa ordered his troops to advance toward Torreón: "His action was hastened by the interception of a wireless message from Pres. Huerta at Mexico City to Gen. Refugio Velasco, commanding the federal garrison at Torreón. As caught by Villa's radio receiver, the message directed Velasco to take the offensive immediately against the Constitutionalists."[54] Consequently, Villa set in motion his own advance and sent information about his actions, including radio interceptions,

FIG. 6. Francisco "Pancho" Villa and some of his men, c. 1914. Courtesy of Library of Congress, Prints and Photographs Division, Reproduction No. LC-DIG-ggbain-10234.

to the U.S. press via imbedded reporters and telegraph messages. The benefits that Huerta hoped to gain from radio communications, in this instance, were countered by wireless's military weakness: it could be overheard. Federal forces in Torreón, however, remained in contact with Mexico City leaders until their complete defeat on April 6, 1914. Because of radio, the Huerta leadership knew well of their loss, but they attempted to spread misinformation to the press and emissaries in the capital until the regime could no longer hide the reality of the situation.[55]

Radio among the Conventionalists

Despite Huertista misinformation to the press and the people of Mexico City, Torreón fell. It was the beginning of the end for the Huerta regime. The general left for exile in July as his remaining forces continued to deteriorate. But instead of ending the fighting, the violence escalated further as the Constitutionalists split, starting another round of civil war. To work out differences and settle the post-Huerta state, the various factions within the Constitutionalist coalition met in the city of Aguascalientes in October 1914. Highly contentious from the outset, it failed miserably, ending any dreams of peace. Instead of mending differences, the meeting facilitated the joining of Villista and Zapatista forces under the "presidency" of convention-elected Eulalio Gutiérrez against those who remained loyal to Carranza. The former group designated themselves the Conventionalists, while the later retained the name of Constitutionalists. Almost immediately, the Conventionalists captured Mexico City while the Constitutionalists fled to Veracruz to set up their base of operations. However, as it quickly became apparent that Gutiérrez was little more than a "storefront" president, as one historian put it, Villa and Zapata retained their positions of leadership and fought the Carrancistas on their own terms.[56]

Once in nominal control of Mexico City, the Gutiérrez administration worked with the remaining SCOP officials to expand the power of the Chapultepec station beginning in November 1914.[57] They also took time to repair the telegraph lines connecting their bases of support in Morelos and to the north. The relationship between the Conventionalist and SCOP leadership, however, never had time to cement. Gutiérrez, members of his general staff, and the Villista and Zapatista armies, possessed a limited understanding of how the Mexico City bureaucracy worked.[58] The civil war itself disrupted any chance of educating themselves. Before the improvements to the Chapultepec station were complete, Constitutionalist forces were already on the march to retake the capital.

Throughout much of 1915, the fractious rebel forces fought over Mexico City. After Carrancistas took control of the capital in January, the first time they controlled the city since their original evacuation following the Aguascalientes Convention, Villa moved back to the north and Zapata to the south, the regions where each felt most secure. Conventionalists continued to force Carrancista generals Obregón and González to evacuate the city after securing it for short periods of time, the latter for only eight days in July. Indeed, the capital changed hands five times from July 1914 to July 1915. On each occasion the surrendering army sabotaged the wireless station and then the invading forces worked to repair it.[59]

By February 1915, Zapatistas possessed a radio operation in Cuernavaca named the "Clandestine Wireless Office."[60] The beginning date for this station—at least when it shows up in historical documentation—suggests that the equipment arrived with the provisional Conventionalist government as it fled to Cuernavaca from Mexico City in late January 1915. L. G. González and a handful of other operators used the office's receiver to intercept messages.

What they found varied greatly. Transmissions ranged from personal regards to family members to information about Constitutionalist operations to messages sent to business headquarters from U.S. shrimp companies. Most overheard transmissions came from Carrancista stations along both the Gulf of Mexico and the Pacific Ocean, which were frequent. Often these messages were about small financial matters and the movements of individuals or troops. Many of them were inconsequential, but others consisted of discussions on important topics by Constitutionalist leaders including Luis Cabrera, Governor Luis Caballero, General A. I. Villarreal, General Jesús Acuña, General Álvaro Obregón, and even Venustiano Carranza.

Some stations under Constitutionalist control remained open to the public, and were used for personal communications. Transmissions intercepted by Zapatistas from February 28 included a message from Tampico to Veracruz from a Mrs. Cármen Téllez stating that she was leaving and hoped that her family would wait for her. "Give my love to the children."[61] Other messages were about domestic struggles in homes where men had gone to war. One woman on March 1 pleaded for her husband to send her money, as she had none left. A wireless telegraph to a young man advised him that his mom had died. Radio not only served as a military tool, but also as a rare connection for some soldiers to their homes and noncombatants living though the difficulties of war.[62]

Many of the military messages were purposefully vague or cryptic. Most army officers had learned that they could be overheard. U.S. Navy ships almost always sent their information in code, so too did Carranza, much to the frustration of the Zapatistas. J. Cosío Robles, a Carrancista officer in Campeche, in a message to fellow Constitution-alist P. M. Navarrete, stated "María told me where you are. I want to talk with you."[63] These vague types of messages were common; still, the Clandestine Wireless Office compiled them all in hopes of finding anything useful.

Some messages were of greater military value. An intercepted transmission sent by Carrancista general J. B. Treviño stated that one Mr. Vazquez needed to organize "a troop" and have them "disembark" soon. On March 3 the Cuernavaca office listened in to a fight between Constitutionalist Captain C. A. Molina and Colonel C. C. Martínez over the loyalty of Molina's troops. A "circular" to top Carrancista leaders on March 8 discussed the defeat of Villista forces near San Juan del Río, Querétaro.[64] With these interceptions, Zapatistas gained some insight on troop movements, defeats and victories, supplies, and enemy morale.

Although interference made reception difficult at times, the equip-ment in Cuernavaca generally functioned well. The intelligence gatherers worked at night, when reception was clearest. According to González, "the results in the wireless office were far better than I had originally hoped for."[65] One night he picked up coded messages from a radio in Zapatista territory, an unknown portable device sending messages to the enemy. During the same listening session he tuned into transmissions from a Japanese vessel on the high sea north of the Isla Marias.[66] How surreal it must have felt to have been dialing in mysterious signals from far away while in the throes of a struggle for existence at home.

Foreign businesses and ships provided another source of messages. Companies based out of San Francisco and New Orleans communicated regularly with ships off Mexican waters and with the Mazatlán and Vera-cruz stations. Shrimp boats worked off the coast of the Baja California

Peninsula. Foreign oil companies in Mexico, such as Waters-Pierce and El Aguila, used radio to keep business going.[67] Most, though not all, of these businesses were engaged in trade with Mexican companies and with Constitutionalist forces, all while the U.S. navy anchored nearby. U.S. businesses were crucial to Carranza. They provided food, military supplies, and information from the outside world. To Carranza's enemies, messages from these ships provided information on some of the First Chief's resources and alliances.

The Zapatistas also possessed at least one field radio. They used it in the battles over Mexico City, likely for communicating between the front lines and Cuernavaca. In July 1915, Carrancista General Coss captured one such Zapatista unit retreating from Mexico City.[68] The predominate image of Zapatistas as proud but homespun peasant rebels remains, but they achieved a certain degree of technological sophistication little acknowledged.

Villa built his radio operations on more established foundations. He relied substantially on electronic communications, including radio, before and after his brief stint in Mexico City. Radio played a key part in his attempt at economic stimulation in his Chihuahua stronghold. Silvestre Terrazas, Villa's close aide, wrote, "With the wireless station, we achieved immediate communication with the whole world. During Villa's occupation of Chihuahua, the area immediately witnessed the consolidation of all kinds of communication, and with the well cared for postal service, greatly facilitated activity and transactions throughout the region."[69] Other sources confirm Villa's efforts to establish a well-functioning government and economy in the region.[70] A former part of Carranza's war machine, Villa had experience in using radio to build local economies and transnational alliances.

Villa built other stations in areas he controlled in northern Mexico as well. In May 1915, his men constructed a radio office in the city of Durango. Villistas built another station in Ciudad Juárez. Villa's forces also possessed wireless offices in the city of Chihuahua, Torreón, and in the field. His communications system was impressive, but starting that year, the engineer-laden Constitutionalist forces would take over his stations one by one.

Conclusion: Radio, Civil War, and U.S. Intervention

A review of the first five years of the Revolution makes it clear that the upheaval increased the presence of wireless technology while shifting radio use from state expansion to the needs of factional combatants. Under Madero a number of important advances were made, including wireless communication with foreign nations, but development occurred largely along the lines already established during the Porfiriato. Madero, and especially Huerta, increased radio imports specifically for military purposes. Following the assassination of Madero, Huerta's opposition, the Constitutionalists, built their own radio network in response to the overthrow, ultimately benefiting from U.S president Wilson's distaste for Huerta. When Carranza could no longer maintain control over his own armies, the breakaway Conventionalists used their portion of wireless equipment to facilitate governance and to gather intelligence.

However, as discussed in chapter 3, it was the Constitutionalists who remained loyal to the First Chief who proved most adept at communications. The struggle to establish dominance in wire and wireless telegraphy had become an essential component of military and administrative planning. The Carrancistas were well aware that as long as any side failed to monopolize communications, the struggle would continue to divide and damage the nation. Carranza used a close cadre of engineers to help him solidify control over information networks, including radio stations.

The operations of the U.S. navy and diplomatic core also proved disjointed but significant to the course of the war. Whatever the motives of Wilson's decision to oust Huerta, and there were many, enacting a clear policy in Mexico proved difficult.[71] The reasons were twofold: first, Wilson faced a barrage of contradictory accounts from diplomats, special agents, and admirals; and secondly, the actions of naval personnel along the Mexican coasts complicated matters. While the newly formed

Wilson administration strove diligently to understand the situation, naval officers in different theaters often acted on their own accord. Nonetheless, Wilson's dislike of Huerta ultimately favored Carranza, especially after Huerta dissolved Congress on October 10, 1914, and Wilson more firmly set his mind to regime change in Mexico.

The Constitutionalist forces gathered steam as they gained greater access to American supplies and as the Huerta administration suffered from constant U.S. harassment. One of the more important military imports from their northern neighbors, radio proved crucial to Carranza's operations, and the Constitutionalists utilized the equipment with skill, helping to facilitate victories over Huerta, and again during the renewed civil war against Zapata and Villa.

3   Rebuilding a Nation at War

Wireless communications defied the bandits, and in a large measure
our success was due to our ability to keep in touch with centers of
supplies and the outer world.

Ignacio Bonillas, Constitutionalist Minister of Communications, 1916

From the port of Veracruz, Venustiano Carranza plotted the demise of
his adversaries and the reconstruction of Mexico. Welcomed with great
fanfare, the Carrancistas strained the local economy but improved the
city's waterworks, markets, and public spaces.[1] Even if most Carrancistas were not local, at least they were not Americans. The U.S navy
had just ended its occupation of the city. To facilitate the movement
of Constitutionalist forces and to maintain a consistent link with his
generals in the field, the First Chief used the Veracruz radio office and
sent out loyal technicians to rebuild other stations. The U.S. military,
in its hasty departure in November 1914, left warehouses of military
equipment, including artillery and field radios. After stocking up on
armaments and food, Generals Obregón and González set out after
Generals Villa and Zapata with radios and guns in hand. In Veracruz,
Carranza eagerly awaited news of the capture of Mexico City and the
destruction of his enemies.

After months of battling over the nation's capital, Carranza's forces
entered Mexico City once again on July 31, 1915. Their position was

FIG. 7. Venustiano Carranza and advisers, c. 1913. Courtesy of Library of Congress, Prints and Photographs Division, Reproduction No. LC-DIG-ggbain-14637.

fragile, but they were determined to stay. The taking of the capital, in addition to subsequent victories over Villa's forces at the cities of Celaya and León, allowed Carranza's government to obtain a firmer grasp on the country while marginalizing its enemies. During the next five years, the Constitutionalists strove to bring the whole of the national territory under their control, making great strides though never completely succeeding.

Radio was crucial for reconsolidating state control and obtaining recognition from governments abroad. First used by Carrancistas to challenge Huerta, Carranza's radio experts moved to monopolize wireless communications across the whole of Mexico. Radio proved vital for defeating opposing revolutionaries and reactionaries still at large while bringing nationalist centralization schemes to new heights. In the process, the Constitutionalists passed laws on radiotelegraphy that would become the foundation for broadcasting legislation in the 1920s. Lastly, radio became a common element in foreign relations under Carranza, not only in interactions with belligerent empires fighting World War I, but also in opposing U.S. imperialism and expanding Mexican influence in Central America. These trends, too, would continue in the following decade.

Winning the Civil War

Radiotelegraphy proved essential to Carranza's actions against opposing factions from 1915 to 1920. Warfare had taken a toll on the country's transportation and communication networks, and repairing sabotaged telegraph lines became increasingly difficult and costly. Of course, Carranza ardently repaired these links, but to help offset the problem, the Constitutionalist leadership sought out radios. In April 1917, Carranza controlled sixteen functioning stations—four of which he deemed high power, four wireless sets aboard ships of war, and five portable military radios. Remaining in a volatile situation, Carranza comforted Constitutionalist leaders by highlighting their increasing dominance over radio communications.[2]

Carranza had possessed wireless devices throughout the civil war. The station in Veracruz, the First Chief's temporary base of operations in late 1914 and much of 1915, remained functioning and was a crucial element of Carranza's communications network. Carranza's officers reestablished wireless links between Saltillo and Veracruz. They also maintained control of the high-powered Campeche station and repaired the Tampico transmitter.[3] The gulf stations were especially important to Carranza. They made possible international trade, including oil and henequen, and ongoing conversations with military leaders and outside political agents, increasing Carranza's chances of recognition by foreign governments.

Obregón used wireless devices in his spring 1915 campaign against Villa. Incorporating machine guns, artillery pieces, and radios left behind after the U.S. invasion of Veracruz, Obregón obtained a clear superiority in military technology over Villa.[4] Although it is debatable whether the Americans purposefully left this equipment specifically to aid the Carrancistas, it was left and used nevertheless. On August 3, Obregón radioed Carranza in Veracruz from Salamanca, Guanajuato, with news of Constitutionalist victories in Aguascalientes, San Luis Potosí, Querétaro, and Zacatecas.[5] His forces were pushing further and further into Villista territory.

Carranza had also obtained much of Huerta's small but significant navy, including radio-equipped vessels such as the *Guerrero* and the *Melchor Ocampo*. The latter saw important action in battles over Tampico in 1915 and 1916. Previously, in 1914, the radio-savvy González ordered wireless devices placed aboard the *Zaragoza*, an old corvette. By April 1917, the Constitutionalists had at least four ships with radios. The last of these, the *Progreso*, was a war transport that an American company in New Orleans revamped with "a modern radiotelegraphic station with a range of six hundred miles in all directions."[6] Before Obregón's own rise to power in 1920, wireless specialists additionally equipped the government ship *Chiapas*. This small, radio-equipped navy monitored the shores for enemy movements, defeated opponents, contacted outsiders, and delivered aid. Control of these ships, and the information and goods they provided, gave the Carrancistas an important edge.

Following the Conventionalists' final expulsion from Mexico City, the Villista and Zapatista radio operations fell one by one into the hands of the marching Constitutionalists. In August 1915, Jesús Acuña, Carranza's minister of foreign relations, proclaimed that the Chapultepec station was "now in complete working order."[7] With field radios originally used against Huerta in addition to those left behind by American forces, Obregón and González pushed into enemy territory.[8] By the end of December 1915, Villista officials turned over Juárez and its radio station to Constitutionalist agents. Carrancista general Francisco Murguía took Torreón from the east and Obregón took Chihuahua City the following year, gaining the wireless operations in those respective cities as well. In the summer of 1916 González took over the Zapatista station in Cuernavaca.[9] Piece by piece, Carranza's enemies lost control of their wireless network.

The Constitutionalists also used radio to keep tabs on and interfere with U.S. forces, especially after Villa's raid on Columbus, New Mexico. Desperate after a number of disastrous defeats and angered by U.S. recognition of Carranza, Villa and a small band of his soldiers raided the small American town on March 9, 1916. This action would once again provoke the Woodrow Wilson administration. He sent the Punitive Expedition into northern Mexico, led by General John "Black Jack" Pershing, to destroy the Villistas. Perhsing's forces clashed with Villista and ultimately Carrancista forces in the massive northern state of Chihuahua. The expedition nearly made it to the border of Durango before U.S. government leaders ordered it to return after it nearly caused an all out war with Mexico. In response to the initial invasion, the Constitutionalists shifted their radio use in a new direction—spying on American ground forces. The Carranza government wanted Villa's remaining army disbanded. The First Chief, however, could not look overly tolerant of Americans in Mexico. He definitely did not want to

FIG. 8. U.S. General John J. "Black Jack" Pershing and his aide, Colonia Dublán, Chihuahua, Mexico, 1916. Courtesy of Wikimedia Commons.

look like a puppet of U.S. whims, precisely what Villa accused him of being. Carranza's forces kept tabs on American and Villista movements, using spies and Villa's old border radio station in Juárez.

The American chase for Villa was a wireless event. In fact, it was a major testing ground for U.S. military technologies as a whole. Although the American armed forces had used planes, automobiles, and radios to increase vigilance along the Mexican border, the invasion provided the first application of these technologies during a military campaign in a foreign territory. Despite difficulties, American journalists sent back news of the expedition via wireless to the United States on a daily basis. In the United States, newspapers printed these reports in sensationalized accounts in cities across the country. Soldiers in the field and naval radio operators in warships along the Mexican coasts sent and received messages from American forts along the U.S.-Mexico border. U.S. intelligence officers in Pershing's army also intercepted Carrancista radio messages. Arguing that Carranza's armies could do little to stop Villista forces, Pershing cited a radio message from General Jacinto B. Treviño to Obregón in which Obregón told Treviño that blankets could not be provided to soldiers because a contractor failed to come through.[10] U.S. forces often paid as much attention to the Carrancistas as the Villistas.

Carrancista radio operators, in turn, interfered with and surveilled American transmissions. U.S. army officials regularly complained that their services were "repeatedly interrupted by high power radio waves sent out from one or more [Mexican] stations."[11] Of course, intentional sabotage by Mexicans was not the sole problem. Constitutionalist radio operators sending reports to Mexico City about U.S. messages they overheard caused inadvertent interference. American messages also "got tangled up with such things as wheat quotations" or other nonmilitary-related transmissions.[12]

Carrancista spies were more interested in using radio for gathering intelligence than disrupting services. The Juárez station, freshly taken from Villa, proved especially useful. On a nightly basis operators listened to the messages transmitted to and from Fort Bliss and Fort Sam Houston, Texas; the American army in northern Mexico; and other U.S. stations along the border. Radio listeners in Fort Bliss, Texas, were surprised to discover the Juárez operation when they overheard Mexican radio operators discussing American messages. U.S. military officials thought that Carranza's soldiers had disabled Villa's station in Juárez.[13] They were wrong. In fact, a Carrancista employee had just recently finished slapping a new coat of paint on the facilities in addition to refurbishing the electric wiring.[14] Radio did not likely change the outcome of the Punitive Expedition, but it did help Carranza maintain a rapid system of intelligence, informing the decisions of Constitutionalist leaders fighting a civil war and maneuvering through a destabilizing U.S. intervention.

As the Punitive Expedition plodded its way out of Mexico, Carranza's forces dominated wireless communications across Mexico. Serving as Carranza's minister of communication, Ignacio Bonillas, who would later become Carranza's infelicitous selection for the 1920 presidential succession, stressed the essentialness of radio to the Constitutionalist cause. In early November 1916, he said that as the war progressed and they acquired more territory, communications grew more difficult, leading Carrancista officials to "establish a system of wireless telegraphy that extended all over Mexico. . . . The overwhelming importance of such a system was demonstrated during the revolution. Wireless communications defied the bandits, and in a large measure our success was due to our ability to keep in touch with centers of supplies and the outer world." To obtain these apparatuses, Bonillas furthered the dispatch of Constitutionalist experts to the United States to secure the "most modern appliances and inventions." Nearly in the same breath, he commented that government officials had taken over the wireless operation at the American Copper Company's Cananea mine in Sonora.[15] The Constitutionalists would buy American products, but they would not tolerate private U.S. wireless operations in Mexico, at least not without Carranza's consent.

Publications by Modesto C. Rolland, an engineer and former Carran-cista communications officer who was serving as a Mexican congressman and propaganda agent in the United States, provided comments simi-lar to Bonillas's. In 1917, he argued that "the wireless service has been improved to such an extent that we are able to make the assertion that the entire Republic is covered by stations that control the country in a far more efficient manner, proportionately, than the same service does in the United States."[16] Although a bold and perhaps overstated claim, the Con-stitutionalists had indeed made significant headway in expanding wireless communications across the nation. This expansion was absolutely critical to their military successes and their subsequent attempts at nation building.

Nevertheless, Villa, Zapata, and Félix Díaz (the nephew of Porfirio Díaz involved in the initial overthrow of Madero) remained threats to Carrancista control. Villa, for example, forced Carranza to maintain a large military presence in Chihuahua and Coahuila where Villista guerrillas remained active.[17] The stations in those states continued to be used for military operations. Indeed, communications officials updated these stations with new high-powered devices. Still, Villa briefly took Torreón, the home to one of these facilities. Carranza's officers, including Francisco Murguía, who repulsed Villa from the city of Chihuahua in 1917, possessed "two mobile devices that communicate with the wireless stations of Chihuahua, Tampico, Alamos, etc . . . which has been improv-ing the condition of our forces in that region."[18] Radio was a regular component of intelligence gathering in the last campaign against Villa from 1917 to 1920. But despite the technological edge, the Carrancistas never caught Villa, who prided himself on his intimate knowledge of the northern mountains.

The Zapatistas also remained a serious if weakened threat. General Pablo González had revamped the newly acquired wireless office in Cuernavaca in 1916, and the Carrancistas had taken possession of most, if not all, of the Zapatista radio equipment. Meanwhile, the forces under González's command ravaged the countryside. Still, by the end of the year, González had failed to suppress the rebels who continued to wreak havoc on trains and along the southern fringe of Mexico City. He subsequently withdrew from the state without success. It is not clear if the Zapatistas recovered any of the wireless equipment afterward. It appears that González took it with him. Morelos remained largely in Zapatista hands until 1918, when González marched into Morelos once again. Even after the assassination of Zapata in 1919, many of Morelos's residents resisted Constitutionalist control until Obregón gained power in 1920, thereafter making amends with much of the remaining Zapatista leadership.[19]

Carranza considered the reactionary Félix Díaz one of his greatest threats by 1917.[20] Díaz used unrest fueled by provincial uprisings against Carranza's policies to gain supporters in central and southern Mexico beginning in 1916. They proposed a return to the 1857 Constitution and defended against what locals perceived as northern interlopers. Even without Díaz, states including Tabasco and Chiapas had large populations that resisted Carrancista forces. Díaz took advantage of these sentiments.

Díaz's forces prompted Carranza to expand radio communications into Tabasco. In an attempt to establish greater control over the area, Carrancista SCOP officials started construction on a radio tower in Villahermosa in 1917. Some within the SCOP had wanted to build the wireless office in the coastal town of Frontera, but Governor Luis Domínguez argued that Villahermosa was the better choice because it was the state's capital, centrally located, and would be better situated to assist against Díaz's operations near the Tabasco towns of Teapa and Tacotalpa.[21] The Felicista forces that operated through Veracruz, Oaxaca, Chiapas, and the border of Tabasco jeopardized communications linking Mexico City with parts of the Gulf Coast, the Isthmus of Tehuantepec, and Yucatán.[22] Building the station in Villahermosa allowed for a consistent communications link between these areas without relying on the always susceptible wire telegraphs.

Throughout Carranza's rule, the military continued to expand radio. In 1918 and 1919, the government built stations in Guadalajara, Jalisco, and Córdoba, Veracruz, both "dedicated to the military necessities of the states, in combination with two portable stations."[23] They used these instruments to connect urban bases of operation to combat zones and to counter the sabotaging of telegraph lines in contested areas. Wireless technology helped the Constitutionalists keep tabs on American, Villista, Zapatista, and Felicista forces, which allowed Carranza to maintain at least a modicum of control over large swaths of territory.

Rebuilding the Nation

The headline to a March 12, 1917, article in the pro-Carranza newspaper *El Pueblo* read "One of the Great Works of the Revolution." Discussing the growing wireless system, the article stated that "the services provided by this network to date are immense, and are due, in large part, to the progressive march that has followed the military operations in all of the national territory."[24] State building and conquest went hand in hand. Carranza not only sought out military success but also wanted to establish the foundations of a new Mexican nation. A number of his administrators pushed for the expansion of radio as a means to modernize and solidify the Mexico of the future, a Mexico they hoped to shape and control.

Indeed, radio development became a crucial part of a renewed nation-building campaign. Although nationalist tendencies were not new when the Revolution broke out, Carranza brought rhetoric about Mexican sovereignty and nationalism to a new fever pitch. Carranza further nationalized the railway system, taking control of the lines that had remained in foreign hands. He also proclaimed state control over the railroad telegraph and telephone operations. By mid-1917 the SCOP took up a number of public works projects as well. These were mostly enterprises started during the Porfiriato that had been abandoned during the Revolution, including new railroads, harbor improvements, irrigation development, and the national theater in Mexico City.[25]

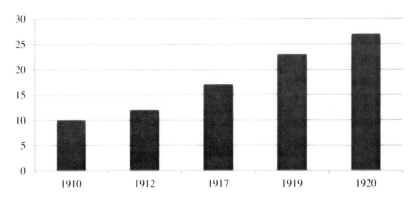

FIG. 9. Functioning state radiotelegraph stations, 1910–20. Radiotelegraph stations more than doubled during the Revolution. (This graph does not include mobile or foreign radio stations.) *Sources*: Merchán Escalante, *Telecomunicaciones*, 63; "Una de las grandes obras de la revolución," *El Pueblo*, March 12, 1917, 10; Venustiano Carranza, in *Los presidentes de México ante la nación*, 357; Adolfo de la Huerta, in *Los presidentes de México ante la nación*, 409.

Carranza's government enacted two important laws on radio. Government policy since the Porfiriato had been that all wireless operations obtain a state permit to operate legally within the country, but most of the U.S. mining businesses possessing wireless equipment failed to register their devices. Porfirian officials had either not known about or ignored these trespasses, but the First Chief shut down these stations. For one, their existence provoked his nationalist sentimentality, but more importantly, they sent messages about Mexican forces to the U.S. military, consulates, and businessmen. In mid-March Carranza established strict censorship on these stations. Future president Plutarco Elías Calles, then military governor of Sonora, ordered the dismantling of the wireless plants in the Cananea and Nacozari mines.[26]

On October 19, 1916, in response to the continuing civil war but also to the Pershing expedition, Carranza and Manuel Rodríguez Gutiérrez—Bonillas's successor as minister of communications—published a decree clarifying the First Chief's policies on radio. The regulations reiterated that individuals or companies could not operate radio equipment without government permission. They also deemed illegal any divulgence of government information transmitted via radio. Those who broke the law could face a 500 to 1,000 peso fine, up to eleven months in jail, and the confiscation of their equipment "to the nation."[27] A difficult law to enforce, but it proved useful in providing a legal basis for expropriating enemy radio equipment. This law affected foreigners in addition to local enemies because only two of the dozen or so foreign radio offices in Mexico were legally registered.

Legislators further codified wireless communications in Article 28 of the 1917 Constitution, which disallowed private or government monopolies excepting those "relating to the coinage of money, to the postal, telegraphic and radiotelegraphic services."[28] Carranza believed that this section of the document held significance because it restricted the creation of foreign monopolies.[29] It also clearly categorized radiotelegraphy as a type of public work to be controlled by the government. To help enforce the new policies, Carranza sent "special wireless men" to search "the hills near Mexico City, and [to make sure] all state governors have received instructions from Mexico City to keep a close watch to prevent the establishment of wireless plants without government authorization."[30] This action also served to counter American claims of rogue German operations in Mexico; after all, World War I was now in full swing, a topic discussed later in this chapter.

In the process of building a viable wireless network to combat enemies, the Carranza administration renewed the old goal of increasing central control over Baja California and Quintana Roo. In late February 1917, the business paper *El Economista* stressed that radio was a "great benefit to all the peninsula states because they can obtain rapid communication with the interior of the country."[31] Employees of the DGTN erected a high-powered station in Salina Cruz, Oaxaca, and started new works at Puerto Morelos, Quintana Roo, and Bahía Magdalena, Baja California. They repaired and refurbished the station at Xcalak, which a hurricane had destroyed. Workers fixed the device quickly because "the remote area" had "no other kind of communication with the rest of the country."[32] In 1919, the government finished building another high-power station in La Paz, Southern District of Baja California, to replace the San José del Cabo office, which had also been knocked out by a hurricane.

The return to a frontier-connecting radio policy is further demonstrated in Rolland's continued writings. The same year that the government finished the La Paz station, he headed a commission designed to report on the economic and infrastructural conditions in the Northern District of Baja California. Another underlying goal of the mission was to bring the Baja California Peninsula, which had remained largely independent during the Revolution, back under federal control. Rolland believed that one of the best means of establishing a stronger presence was wireless communications. He praised the region's radio-telegraphic operations. Carranza officials had recently built powerful stations in La Paz in the south, and in Mexicali, the capital of the north. Constitutionalists also took possession of local stations built under the watch of renegade governor Estaban Cantú in Ensenada, Tijuana, Tecate, and Algodones. These wireless outposts not only linked the territory together, but through Mexicali "intimately united this remote region with the center of the republic that had lamentably lived disconnected due to a lack of efficient communication."[33]

Rolland additionally warned that northern Baja California had become too reliant on the United States to relay information. This situation became especially problematic when American officials placed stricter regulations on communications during World War I. Rolland argued that without the Mexicali station, the Northern District of Baja California "would become almost completely isolated due to the severity of the measure employed by the neighboring nation."[34] These stations helped facilitate a long process of attempting, yet again, to drag the peninsula back into the federal system.

World War I and Constitutionalist Foreign Relations

Connecting the frontier peninsulas to Mexico City via radio was a return to the policies of Díaz and Madero, but Carranza went a step further by overseeing the construction of high-power stations in the capital and expanding radio's role in foreign relations. Mexico renewed its participation in international conferences about wireless policies and donated a radio station to El Salvador, attempting to counter U.S. influence in the region. These foreign policy endeavors, however, were aimed just as much at securing Constitutionalist control in Mexico as at asserting their power outside of the country; they were mutually reinforcing undertakings. World War I greatly influenced these developments. In fact, the war, like the Revolution, was responsible for quickly expanding state radio services in Mexico from 1916 to 1919.

German officials took up an even stronger interest in Mexico because of changes in U.S. policies. The German government had relied mostly on the Sayville radio station in New York as their main receiver in the Western Hemisphere for their Nauen transmitter. The U.S. government also allowed Germans access to overseas cables, which they used to communicate with Latin America. In 1915, however, the American navy began listening in on the Sayville messages, and in April 1917, when the United States joined World War I, the American link was completely closed to Germany.[35] U.S. officials additionally began a widespread crackdown on German operatives.[36] As a result, many German radio operators in the United States fled to Mexico.

German agents saw Mexico as the most optimal location for reestablishing a communication link to Germany and a base of operations to spy on the United States and spread German propaganda in Latin America. Indeed, plans for helping the Carranza government build high-power radio stations in Mexico City had been under discussion since 1916. The well-established relationship between Germany and Mexico in radio development and Carranza's well-known nationalist and anti-American sentiments aided these German initiatives. It had mostly been the tentative backing of Huerta by German leaders that caused the Carranza administration to turn earlier to the United States for wireless devices in 1914 and 1915. Carranza's decision to allow Germans to build the Mexico City stations was a return to the long-standing relationship with Germany in all things radio. Carranza was also a practical man. The stations would dramatically expand the government's reach.

It was domestic affairs, namely defeating his enemies within Mexico and securing control over the country, that motivated Carranza to allow the Germans to design the Mexico City stations, which the Mexican and German governments both funded. Despite the fact that Mexican officials willingly transmitted and relayed German messages and propaganda, it was largely Mexicans that operated the transmitters and receivers, which were mostly used for Constitutionalist designs.

It was within this context that the foreign secretary of the German empire, Arthur Zimmermann, sent the now famous "Zimmermann telegram," proposing a military alliance between Germany and Mexico, to the German ambassador in Mexico, Heinrich von Eckardt.[37] The German government genuinely pursued a policy of persuading Mexico to join the war on their side, but Carranza was unwilling to do so. He only seriously considered the notion in the case of a full-out U.S. invasion, a conflict that his advisers worked tirelessly to avoid. Carranza's cooperation with Germans in matters of radio should not be mistaken for taking the side of Germany in World War I. Carranza remained a proponent of neutrality.

Von Eckardt headed a number of German operations in Mexico, including the construction of the massive radio plants. Beginning in 1916, Von Eckardt solicited his superiors in Germany for the creation of a high-power wireless operation in Mexico City, arguing that it "would make us independent of the North American stations."[38] The German Reich's Post Office originally resisted the idea, stating that the greater distance and cost made the station undesirable. Its leaders backed down, however, in the face of growing communications problems and increased pressure from the German Admiral Staff and Foreign Office, both of which supported the idea. The Germans planned to bring the equipment for the station from New York, where a Telefunken subsidiary still operated.[39] Possessing their own internal motives for the station, leaders of the Mexican Ministry of Foreign Relations showed their approval for the plan in December 1916. Discussions on the topic, however, remained ongoing throughout the following two months while German agents were gathering the equipment in New York.

Exhibiting the importance of the project, a dialogue about the stations between Von Eckardt and SCOP Secretary Cándido Aguilar occurred regularly, even in the midst of the constitutional convention in Querétaro.[40] Von Eckardt typed his letters with a sense of urgency, almost desperation. Germany, he wrote, needed to keep a communication channel open between Nauen, Germany, and North America. Von Eckardt and Aguilar debated the best methods of transporting equipment from the United States to Mexico to avoid U.S. agents who planned to confiscate the materials. The two officials also wrote about designs to have a German specialist improve Mexican workshops producing radio parts that function well with Telefunken devices. Cost was another important issue. The Carranza administration, via Aguilar, agreed to pay the equivalent of 500,000 U.S. dollars, half of the cost of the proposed station. They Mexican government would, according to Aguilar, let the Germans help manage the plant and communicate with Nauen, but Aguilar reminded Von Eckardt that Carranza had decreed that "only the government of Mexico could establish radiotelegraph stations in the national territory."[41] The Germans would need Mexican approval, and they would have to bow to Mexican needs before they could get theirs.

While working out the details about the Mexico City station, the Carranza administration, with the assistance of German specialists, donated a wireless station to El Salvador in January 1917. The equipment was of German and Mexican make. Led by the ever-important Luis Sánchez, the same man who had built equipment for Díaz and Huerta, the team of Mexican specialists built the four-kilowatt station and trained Salvadoran telegraphers. Via the Mexican warship *Jesús Carranza,* the Mexican communications officers arrived along with towers and other necessary parts.[42] Building radio stations in Central America had actually been a goal of Mexican state officials since 1912, when they had started their first international broadcasts and campaigns to extend telegraph and telephone service abroad.[43] Interrupted by the Revolution, Sánchez and company did not complete the tower and connecting office until September 1917.

The renewed emphasis on the Salvadoran station resulted from three intertwined foreign-relations issues: Carranza's attempt to expand Mexican influence in Central America, El Salvador and Mexico's joint opposition to American and Guatemalan power, and the renewed communications relationship between Germany and Mexico. Although *El Pueblo* proudly proclaimed that Carranza was establishing a new "vigorous and strong" policy in Central America, his administration built squarely on Porfirian foundations. President Díaz had possessed firm relations with Salvadoran leaders, and he had hoped to raise his international prestige as a mediator in Central American affairs. He additionally tried to counter the influence of the United States and Guatemala in the early 1900s.[44] By the time Carranza had obtained the presidency, El Salvador was the only Central American nation that remained truly friendly to Mexico. And Carranza, like Díaz before him, believed that Mexico was an example for Central Americans to follow.[45] Even more than the old dictator, Carranza presented himself forcefully as the torchbearer of nationalism and the protector of sovereignty in Latin America.

Carranza and many of his administrators lambasted American imperialism and all that it included: dollar and gunboat diplomacy, and U.S. military intervention in Latin America.[46] Carranza, playing on Wilson's proclamations about national self-determination, wrote editorials promoting nationalism and the right to self-rule without U.S. interference. He appealed not only to the revolutionaries of Mexico, but to the "revolutionaries of Latin America, the revolutionaries of the universe."[47] Sharing a large border with the United States, Mexican officials and artists showed Mexico as leading the way against Yankee imperialism. Carranza hoped that the Salvadoran station would help spread this message to Central America, counter U.S. dominance, and keep an ear on Guatemala's legendarily paranoiac strongman and U.S. ally, "El Señor Presidente" Manuel Estrada Cabrera.[48]

There were other important motivations involved in building the station as well. Confronted with inconsistent trade and embargos from the United States, El Salvador became a reliable if limited partner for the Constitutionalist forces. In exchange for the radio equipment, two Mexican-constructed aircrafts, and assistance during a devastating earthquake, Carranza received good deals on ammunition and a reliable ally on Mexico's southern border.[49] El Salvador was also the only Central American country that, like Mexico, took a neutral stance during World War I. German operatives full-heartedly supported the radio project there and helped provide the necessary equipment and expertise. By 1918 El Salvador received daily bulletins from Mexico, including German news and propaganda.[50]

While Sánchez raised the towers in San Salvador, German engineers and other SCOP officials got the Mexico City project off the ground. Construction began in March 1917. The shipment of radio materials had left New York on February 28, the day before the U.S. press published the soon-to-be notorious Zimmermann telegram.[51] Historian Friedrich Katz hypothesized that the publicity surrounding the telegram caused American naval officials to halt the ship and to confiscate most of the materials before the crew could unload the goods in Veracruz. Katz's assumption seems reasonable. The public release of the Zimmermann telegraph caused an uproar in the United States and greatly heightened American sensitivity to German communications in the Western Hemisphere. However, the letters from Von Eckardt to Aguilar in January 1917—before the publication of the Zimmermann telegram—show that German operatives already suspected that Americans might try to halt the shipment.[52]

This setback did not stop the construction of the station. Germany had sent other equipment in December 1916, and some of the February 1917 shipment may have made it through as well. German and Mexican agents also did a good job of improvising in response to the circumstances. American officials had been mostly correct to note Mexico's reliance on foreign technologies, but the increased difficulties in obtaining war materials because of U.S. restrictions and World War I provoked the Carranza administration to become more self-sufficient. One result was that the Constitutionalist military began manufacturing ammunition, airplanes, and communications equipment. Mexican technical groups began to build radio components based on Telefunken designs. Carranza also tapped domestic manufacturers. The Compañia Fundidora de Fierro y Acero de Monterrey provided the iron for the project. A generator and rotor was constructed from parts made in Monterrey and Mexico City. Mexico's National School of Telegraphy also began to add new classes on radiotelegraphy.[53]

From 1917 to 1919, German and Mexican officials worked together to build a receiver and transmitter, which they decided to put in different locations. They constructed the former in Ixtapalapa and the latter in Chapultepec, tackling the receiver first. They chose to build it in Ixtapalapa, a neighborhood on the southern outskirts of Mexico City, to hide it from American intelligence officers but also for technical reasons. The first messages from Nauen arrived in late April 1917. Afterward, Von Eckardt and other members of the German legation and intelligence community received messages from Germany on a regular basis, relayed by Carranza officials sympathetic to their cause. To build the transmitter, the German government relied heavily on Gustavo Reuthe, an engineer who had worked at the Sayville station until forced out of the United States by American officials. Reuthe, Eugene Dzinzilewsky—another German engineer—and Mexican specialists Luis Sánchez, Agustín Flores, Ignacio Galindo, and Salvador Teyabas, completed the station in July 1918.[54] From this point until his death, Reuthe would become an important communications

FIG. 10. Chapultepec radio station, c. 1920. *Memoria por el SCOP*, 1928–29.

operator, teacher, and government consultant in Mexico. He had a huge influence on a number of Mexico's top radio specialists. He also continued to work on Telefunken's behalf and to aid Mexican efforts to expand radio into Central America. But the Germans never completely controlled the Mexico City stations. Early transmissions also failed to reach Nauen. This problem persisted throughout the year, though operators somewhat circumnavigated the issue by sending messages to Spain, which were then forwarded to Germany.[55]

The station was impressive. Consisting of three giant 165-meter tall towers, the 200-kilowatt transmitter could send messages over a distance of 9,320 miles, at least on a good night. It compared well with the most advanced transmitters in Europe and the United States. In tandem with the sensitive receiver that picked up messages from around the world at the Ixtapalapa station, communications officers could send messages to a large portion of Europe, Japan, and South America.[56]

It is essential to understand that German leaders saw the Mexico City radio stations as serving a greater purpose than the reception and

FIG. 11. Workers in the Chapultepec radio workshop, c. 1927. *Memoria por el SCOP*, 1927–28.

transmission of military messages during World War I. By the time the transmitter was completed, German leaders realized they would likely lose the war. They were looking to the future. They feared that they might lose control of their country's international cable telegraphs upon defeat. Radio became a significant component of their plans to continue operating strategic global communications, both for political and commercial reasons. In Latin America, German officials specifically targeted Mexico, Brazil, Uruguay, Chile, and Argentina as potential allies and markets. Reuthe remained an important adviser in Mexico City, and Telefunken remained a large supplier of radio equipment to Mexico and Latin America well into the 1920s, as it had been previously during the Díaz and Huerta administrations.

In the United States, the Zimmermann telegram and subsequent articles about German spies brought fears about wireless communications in Mexico to a near state of hysteria. On March 1, 1917, when the contents of the telegram began to circulate widely among American newspapers, articles described "100,000 Germans" in Mexico, as if they were poised to invade.[57] The same writings described extensive secret wireless networks. One report stated that "authorities" in the San Diego area had known about German radio operations in Baja California for weeks. These included "three wireless stations 600 or 700 miles south of San Diego in the vicinity of San Quintin, Baja Cal."[58] These wireless operations, far from Mexico City and close to the United States, supposedly had collapsible masts that Germans lowered during the day and raised during the night. The *New York Times* further claimed that there was another Baja California station around Turtle Bay, which was high powered and had been overheard "sending in a slow, methodical way."[59] Paranoia was rampant.

One article, published March 9 in the *New York Times,* is particularly interesting for its curious blend of accurate information and critical mistakes. It stated that "semi-official" sources told of a powerful wireless station built in Mexico City that allowed "direct communication between Germany and Mexico City." It also mentioned that one source stated that it was not powerful enough to send messages to Germany, but that it could receive transmissions from Nauen.[60] The strange thing is that the Mexican receiving station in Ixtapalapa was not in operating order until April. The article appears to have caught the German Admiral Staff by surprise.[61] Katz states that "the premature American announcement was either the result of false information by the U.S. intelligence agencies or, since the announcement was made one week after the publication of the Zimmermann note, was consciously calculated to intensify the impact of the note by attempting to exaggerate German influence in Mexico."[62] The details of the newspaper account strongly suggest a U.S. intelligence source, but at least one U.S. operative working with the American embassy in Mexico, someone who had closely monitored Mexican radio development, believed that there was "no plant in the country capable of being used by Germany in communicating directly to Germany, and none can be established without our knowing about it."[63] Perhaps the U.S. government provided the statement to increase

American fears and perhaps to let German leaders know that U.S. intelligence agents were on to their plans for wireless in Mexico, even if they did not know the exact details.

The article additionally discussed the El Salvador station built "just over the Guatemalan border" in Acajutla. The source contended that Germans wanted to influence Mexican foreign policy in Central American and to make "[El] Salvador a base of operations for the invasion of Nicaragua by revolutionists" who would allow Germany to build a canal through the country. This statement played into American fears about losing control over the trans-isthmian route between the Atlantic and Pacific Oceans. Like U.S. intelligence reports, journalists gave little credit to Mexican specialists, arguing that Germans ran the El Salvador station and that it was German agents that brought the equipment there on the Mexican ship *Jesús Carranza*.[64] Although Germans played an important role in training and assisting Mexicans, it was the latter who predominately constructed the station in El Salvador, even if the tower met with the approval of Germans. And although the leadership in Germany welcomed all of these developments, it put little actual effort into fomenting revolution in Central America.

Members of the Carranza administration and the Mexican press were actually divided between pro-ally and pro-German leanings. Pro-German officials in the Carranza administration included Mario Méndez, Cándido Aguilar, Benjamin Hill, and Rafael Zubarán Capmany. Pro-ally leaders included Félix Palavicini, Pastor Rouaix, and Alberto Pani. Despite obtaining significant aide from the United States, Carranza and much of the military leaned toward Germany, though they were just as much against American imperialism as pro-German.[65] Obregón, Carranza's most powerful general, however, supposedly favored the United States and its allies.[66]

It was in this superheated atmosphere that some prominent Mexicans came to the United States and spoke of German wireless outposts and corrupt Mexican officials. Juan Suárez—half-brother of José Maria Pino Suárez, the Mexican vice president whom Huertistas had assassinated along with Madero, and an editor for the pro-ally Mexico City paper *El Universal*—told American reporters and businessmen that even though Carranza was trying to curb German influence, he could be doing a better job. According to Suárez, the greatest threat to the peace between the United States and Mexico was a massive network of German spies in Mexico, especially along the western coast. He raved that they had twenty-one wireless stations, including "the most powerful in existence in the City of Mexico" run by "unscrupulous persons, who would do anything for cash."[67]

Suárez had specific "unscrupulous persons" in mind in his exaggerated statement, including important members of the Carranza administration, men including the Director of National Telegraphs Mario Méndez. Trinidad W. Flores—a telegraph officer and spy for the faction supporting Alvaro Obregón during the 1920 presidential succession—later noted that Méndez admired the Germans, welcoming the officials who had operated the Sayville station in New York to work at Ixtapalapa and Chapultepec. According to Flores, Méndez happily relayed information to Von Eckardt and flaunted German-Mexican connections.[68]

U.S. intelligence gatherers confirmed Mexican involvement in relaying German information in the fall of 1918. By that time, the Chapultepec station could finally transmit across the Atlantic, and American radio operators along the U.S.-Mexican border had been regularly searching the airwaves for suspicious signals coming out of Mexico. The Military Intelligence Division sent reports on their results to the military attaché in the American Embassy in Mexico City, the State Department, the Office of Naval Intelligence, and the Department of Justice. After connecting an "unknown 6300-meter station," which operated each night between 11:30 p.m. and 2 a.m., to Chapultepec, the Military Intelligence Division spied on the frequency from August through October. They reported each week on the number of messages sent in code, Spanish, and English. On October 3, they reported that "there have been developments in the Radio Subsection which may influence our Mexican relations. Evidence has been secured that seems to convict the Mexican government of conniving in the German use of radio at Chapultepec for hostile purposes."[69] Although the United States and Mexico never came to war over German communications in Mexico, the evidence supports claims that certain Mexican leaders helped relay information between Germany and German agents in Mexico.

Overall, Carranza and his advisers approached German advances pragmatically. They knew better than to openly support Germany during the war. They had too many problems of their own within Mexico, and once the United States joined World War I, the Carranza administration only entertained the possibility of an alliance with Germany in case of a full-scale U.S. invasion of Mexico.[70] Even if Carranza resented American interference, he still relied heavily on U.S. supplies, and he had no desire to risk more direct intervention. But he also concluded that allowing Germans to help build powerful wireless stations in Mexico had great domestic advantages.

Mexicans had their own goals for these stations. In exchange for relaying German news and messages, the Carranza administration acquired its most powerful communications tools, advancing its influence over the Mexican nation and its ability to communicate with representatives of other countries. Although German specialists were essential to construct the works, it was usually Mexicans who operated them. These were Mexican stations. Obregón, for example, supported the projects because they provided a huge military advantage, allowing Mexico City to better communicate with armies throughout the country.[71] The new stations in Ixtapalapa and Chapultepec allowed for a more consistent means of sending wireless messages across the whole of Mexico and to receive them in return, and telegraphers used them more for these operations than for communication with Germany.

These stations also fit well with the Carranza administration's foreign policy initiatives. In addition to relaying German news, Mexican officials provided their own nationalist and anti-imperialist propaganda abroad. They sent this information to El Salvador, and they had plans to extend radio towers into other parts of Central America.[72] By February 1919, Mexico regularly sent reports to Buenos Aires, Argentina, and Punta Arenas and Valparaiso, Chile. The station operators also began "transoceanic press" services, which they sent to Spain, Germany, Japan, and ships at sea.[73] Outside of the Americas and Europe, the Carranza administration used the German-engineered stations to establish direct communication with Japan. By late 1918 or early 1919, Japan was, at least occasionally, receiving Mexican radio transmissions. This achievement was important to the Carranza government and to Japan. In response to U.S. embargoes, Carranza had turned to Japan to purchase munitions, equipment to manufacture ammunition, and weapons. Japan, on the other hand, had increased its commercial investments in Mexico. The radio link provided a new means of communication that avoided American channels.[74]

The desire of German leaders to maintain communications channels with the Western Hemisphere by building a radio station in Mexico played into Carranza's designs to use wireless to defeat his enemies, secure the whole of Mexico under his government, and strengthen foreign ties. As long as he could publicly remain neutral and not incite the wrath of the United States, which was entangled in the war in Europe, Carranza could play on German needs to fulfill his own. Carranza's pragmatism, nationalist and anti-U.S. policies, American off-and-on again embargoes, and a history of partnership with Germany drove Constitutionalist decision making. As World War I wound down in November 1918, the Carranza government used the German-designed stations for commercial and foreign relations with Germany, the United States, El Salvador, Argentina, Chile, and Japan, but also for domestic communications.

Carranza's Fall

With the European war over and 1920 on the horizon, the Constitutionalists prepared for their first attempt at presidential succession. Obregón appeared to himself and most others as the clear heir to Carranza's position. He was the most powerful general and, despite missing his left arm, was an able and charismatic speaker. Carranza, however, had grown weary of Obregón, who had supported the more radical wing of Constitutionalists represented by the Liberal Constitutionalist Party. Carranza also hoped to have a "civilian" replacement, one that he could influence. Instead of backing Obregón, the clear favorite, Carranza tried to impose Ignacio Bonillas as the next Mexican leader. As a graduate of the Massachusetts Institute of Technology (MIT) and a capable secretary of communications and ambassador to the United States, Bonillas appealed to Carranza as the type of president that a modern Mexico needed. But Bonillas had little support. Bonillas was not well known among the general populace, whereas people recognized Obregón as a victorious general.

Fearing Obregón's escalating power, Carranza attempted to undermine the general's base of support. The First Chief tried to intimidate and co-opt Obregón supporters, including prominent allies in the general's home state of Sonora. The plan backfired. Sonoran military leaders, including Obregón, Plutarco Elías Calles, and Adolfo de la Huerta revolted on April 23, 1920. They issued the Plan of Agua Prieta and prepared their march on the capital. Although Carranza had made progress in strengthening state power over much of Mexico, his failure to address key agrarian issues, his sometimes hostile attitude toward workers and unions, and his consistent clashes with the United States provided exploitable vulnerabilities. Obregón promised greater land distribution and greater political participation, not to mention a possible path of compromise for the remaining Villista and Zapatista forces. Most of the military joined Obregón, and when other prominent generals including Pablo González joined him as well, Carranza chose to flee to Veracruz in an attempt to once again use the port city as a place to regroup and reconquer the country. His ambitions were never realized; he was assassinated in the small village of Tlaxcalantongo in the mountains of northern Puebla. The Agua Prieta rebels took power after only twenty-seven days in rebellion, making Adolfo de la Huerta the provisional president until Obregón could be officially "elected" in November.[75]

Communications played a crucial and fascinating role in the run-up to the 1920 election and the subsequent Agua Prieta Rebellion.[76] Within the SCOP, as with the Constitutionalist forces in general, officials split into factions: Carrancistas, Obregonistas, and those that "swam in both waters," or mediated between both sides.[77] These agents, in turn, fought an internal propaganda and intelligence war, communicating with foreign countries and carrying out acts of espionage and counterespionage. On June 25, 1919, Trinidad W. Flores wrote to a friend that Carranza loyalists in the "Department of Press within the Secretariat of Foreign Relations communicated by wireless to Santiago, Chile, Buenos Aires, Panama, and El Salvador that general González would propose to general Obregón a pact to respect the results of the elections in order to avoid another revolution." Carranza officials spread other statements and rumors favoring themselves to other parts of Latin America, while Obregonistas within the DGTN reported these statements to their leaders, who it turn sent out their own responses. Mario Méndez, the diehard Carrancista director of the DGTN, also ordered his employees not to accept any messages from Tabasco's interim governor Tomás Garrido Canabal, because he no longer possessed "legitimate authority."[78] As tensions grew, Méndez attempted to censor information sent by Obregón supporters and to fire Obregonista members of his staff.[79] His efforts, however, proved to be in vain as the tide of telegraphers, and most others, turned away from Carranza. When the rebellion started, regional Obregón supporters took control of wireless stations. Factions within the central government, including the telegraphers of the secretary of war and marine, also sided with the Agua Prieta rebels.[80]

A week before Carranza's murder, the Sonoran leaders worked out the details of the provisional government, often over the radio. According to one newspaper article, "In a telegram to the revolutionary junta in Washington, Provisional President De la Huerta said that he had been in communication with Gen. Obregón at Mexico City; that Gen. Obregón had recognized the plan of the provisional government, had subordinated himself to it and was in complete accord with all that had been done and with plans in contemplation for perfecting the provisional government. President De la Huerta remains in Hermosillo, capital of Sonora, and has established communications with Mexico City by wireless."[81] Radio allowed the Agua Prieta insurrectionaries to make decisions about leadership despite being separated by hundreds of miles.

During his hasty flight from the capital, Carranza took weapons, records, and money, but he failed to destroy the radio system, which provided the rebels an important communications link to Sonora and the American press. Some stations, however, including in Alamos, Acapulco, and Chihuahua, "had been interrupted by political events."[82] Most of the wireless network nevertheless remained intact, and the Sonoran victors would greatly capitalize on Carranza's previous expansion of Mexican communications.

Conclusion: Radio Control

During Carranza's presidency, wireless communications became an important component of state control and foreign relations. Radio helped secure the Constitutionalists' victory over internal enemies and to consolidate their power over acquired territory. The Carranza administration's ability to communicate with foreign businesses increased its capability to obtain weapons, munitions, and other supplies. The medium also allowed Carranza to speak and to receive messages from foreign political leaders who provided aid and legitimacy to Carranza's cause.

U.S. businessmen provided many of the Constitutionalists' radio devices at first, but Carranza renewed a closer relationship with German providers from 1917 to 1919. The switch was pragmatic and had to do with political and military realities in Mexico, Germany, and the United States. Carranza was aided in obtaining necessary military supplies, including radios, by U.S. officials who opposed Victoriano Huerta (who bought radio equipment from Germany) and then sided with Carranza over Pancho Villa. However, the U.S. government was uneasy about the continuing violence and attempted to influence Mexican policy by putting in place off-and-on again embargoes on this very same equipment. As relations between the United States and Germany soured during World War I, the Carranza administration made the decision to partner again with German agents desperate to keep open radio communications with Germany and willing to help build what would become Mexico's most powerful stations.

During this period, radio continued to be central to U.S. interventions into Mexico. Indeed, Mexico became a testing ground for American military technology in general. The Pershing expedition allowed the U.S. military to test new automobiles, weapons, aircraft, and, of course, radio. Field operators relayed information from the front lines to Pershing and other commanders within Mexico and in American forts close to the border. Carranza communication agents, for their part, kept a constant ear on U.S. radio messages, often using stations recently acquired from Villista forces. Immediately following the withdrawal of U.S. forces, Carranza and a number of his high-ranking officials, including Bonillas, Calles, and Rodríguez Gutiérrez, worked to close down wireless stations operated by American miners—who were correctly seen as spies—while increasing their purchases of U.S. radio equipment.

World War I, like the Mexican Revolution, escalated radio development in Mexico. German agents designed the massive receiving and transmitting stations in the Federal District, providing intelligence for German operatives and a means for Mexican leaders to spread their own news and propaganda. The stations allowed the Carranza government to build stronger links with the rest of Latin America, the United States, Germany, and Japan. With this equipment, Carrancista telegraphers and state officials made wireless technology an important component of Mexican foreign relations. Radio had become an essential component of governance, a point that Obregón understood well and which would become all too apparent during his own time in power.

4   Growth and Insecurity

Mexico has been bitten by the radio bug.
Los Angeles Times, December 16, 1923

The rise of General Álvaro Obregón brought a fragile peace to Mexico. Adolfo de la Huerta, acting as provisional president in 1920 before Obregón's official presidency began, made a truce with Pancho Villa. Félix Díaz fled into exile the same year. After assuming the presidency, Obregón worked out an alliance with the remaining Zapatistas. Small bands of dissidents remained, but no large armies. Obregón endeavored to create stability, but he faced monumental and often contradictory demands from the popular organizations unleashed by the Revolution and industrialists eager to get back to business. Peace was no easy task. Making things more difficult, the U.S. government refused to recognize the Obregón administration, which had come to power through a military coup.

Aspiring to appeal to Mexican nationalism and U.S. representatives alike, the president and his advisers turned to the 1921 centennial of Mexican independence to provide a visual and audio display of Obregón's new revolutionary Mexico. Honoring Agustín Iturbide's successful separation of Mexico from Spain, conservative residents usually recognized the day more than others. But the Obregón administration seized the moment to put on a lavish display. The government organized regional dances, national artwork, airplane shows, baseball games, bullfights, and a new modern marvel: radio broadcasting. The festivities mythologized Mexico's past but also promoted the nation's more modern future.[1] The event was also a public display of Obregón's own policies and persuasions. He billed himself as the man to finally bring the country together and as someone less flagrantly opposed to American interests than Carranza had been.

The Obregón administration's approach to wireless communications was also multifaceted and aimed at different audiences. He built on the structural progress of his predecessor while watching over the diversification and popularization of the technology, especially after the rise of broadcasting. This new form of radio dramatically revolutionized how people thought about the technology, both those who controlled it and those who listened. But underneath the jazz, national airs, and Mexican folk songs that began to beam from commercial and government towers, wireless technology remained a tool of military suppression, rebellion, and espionage. Old fears still shaped government decision making. These realities influenced radio policies and, ultimately, the nation-building designs of Obregón's and subsequent administrations.

As with many of his policies, his decisions on wireless development stemmed from his pragmatist and capitalist leanings but also from reactions to outside impulses that forced him to look at radio in novel ways. Obregón expanded on Carranza's Central American policy of gifting radio stations and fighting against U.S. communications initiatives in Latin America. At the same time, he strove to increase trade with the United States, including the importation of radio receivers. He allowed private entrepreneurs to take the lead in broadcasting while using the state to protect the growth of the Mexican broadcasting industry. He also encouraged the Secretariat of Public Education (SEP) to establish its own station in hopes of initiating a greater direct presence in broadcasting as the government became more firmly established.

The rebellion of Obregón's former ally, de la Huerta, during the last months of the presidency, however, dramatically shifted the government away from pursuing more democratic radio policies. The insurrection reignited fears about the technology as a destabilizing force. The rebellion had extremely important ramifications for radio development in Mexico. The continued reality of civil war pushed state leaders loyal to Obregón and presidential candidate Plutarco Elías Calles toward more authoritarian practices. They shut down operations opposed to their rule while monopolizing political broadcasting themselves. During the rebellion, CYL, the first commercial broadcasting station owned by Raúl and Luis Azcárraga, allied, somewhat reluctantly, with the government, beginning a tumultuous relationship between the state and the Azcárraga family that has survived into the twenty-first century.

Diversification

The 1921 centennial celebration was the first function to popularize broadcasting in the imaginations of Mexico City residents. The attention resulted largely from displays of new radiotelephone devices, which transmitted the human voice and music. Government officials had first used radiotelephones like walkie-talkies for direct communication. However, a number of aficionados and commercial entrepreneurs in the United States, Argentina, and Mexico began using radiotelephony to transmit musical performances to multiple receivers, creating the new field of broadcasting.

In the event's International Commercial Exposition located in the Legislative Palace, the DGTN oversaw a popular booth filled with radios. Showcasing German-, American-, and Mexican-made equipment, the department provided different examples of the "methods of rapid communication with the rest of the Republic and with foreigners."[2] Surrounded by smaller wireless pieces and cable telephones, a radio set with a giant windmill-looking antenna stood prominently in the center of the display. Operated by DGTN officials Agustín Flores and

José D. Valdovinos, a 1-kilowatt De Forest radiotelephone garnered the most attention. Capable of sending and receiving voice transmissions, Flores and Valdovinos allowed people to speak messages to the president, which he supposedly heard through a similar device in Chapultepec Castle.[3]

Many scholars argue that the first radio broadcast occurred at the centennial celebration.[4] That evening, using machines similar to those at the DGTN exhibit, radio aficionados and brothers Adolfo and Pedro Gómez Fernández transmitted a music show to audiences at the Teatro Ideal and the Teatro Nacional (present-day Palacio de Bellas Artes). Airing singer José Mojica and the young voice of eleven-year-old María de los Ángeles Gómez Camacho, the demonstration was the first historically noted musical performance transmitted via radio to a group audience in Mexico.[5] The Gómez station continued to operate, two hours daily, until January 1922.

The centennial festivities also brought attention to the use of radiotelephony in Mexico's young air force. Wireless communications became a component of pilot training during de la Huerta's provisional presidency in 1920, but radiotelephony did not become a demonstrable fact until after the election of Obregón.[6] The day after the centennial celebration, President Obregón, along with members of his general staff, foreign diplomats, journalists, and military officers, drove to the Balbuena airfield outside the capital. There, he presided over the inauguration of radiotelephone services at the base, as well as at a similar installation in Pachuca, Hidalgo. Air force officials had also equipped a Farman biplane with a radiotelephone. Obregón and his entourage watched Mexican aviators conduct aerial stunts. The president talked via radio to pilot Fernando G. Proal, who was a first-generation graduate from the Aviation Academy and one of the country's most respected aviators. Proal communicated with officials at the Balbuena and Pachuca stations, who also talked with each other. The latter station transmitted the popular revolutionary ballad "Adelita."[7]

The following day, the press reported on radio's increasing role in foreign relations. Rafael Cárdenas Jiménez, the Costa Rican consul to Mexico City, told the newspaper *El Universal* that Obregón "decided to donate powerful wireless stations to Guatemala, Costa Rica, Honduras, El Salvador, and Nicaragua with the goal of establishing better relations with Mexico."[8] Obregón promised to escalate Carranza's radio policies in Central America at the same time that he was overseeing the incorporation of the medium in new ways into Mexican society.

The centennial events exhibited the newest developments in radio. The most important change was the radiotelephone. Although the technology had existed in the first decade of the 1900s, it remained impractical until U.S. inventor Lee de Forest made key improvements in the 1910s.[9] Incorporated into the DGTN by mid-1921, it was thereafter applied to interdepartmental communications, aviation, and naval operations. In the words of one newspaper article, "for the Mexican government the introduction of the modern [wireless] telephone communications system reveals the audacity of their work in the ever-increasing conquest of physical sciences employed in human activities."[10] Belief in the power of science was alive and well, and for many urbanites, radio was one of the newest tools in the story of human progress. Obregón embraced these developments, partly because they were a necessary requirement for continued modernization but also because Obregón—a capitalist and inventor himself—genuinely supported them.

Radio had also expanded in other, less publicized ways. When Obregón took office, twenty-seven radiotelegraph stations operated regularly across the country.[11] The workshops of the DGTN amped up their production of radio equipment, providing receivers for the centennial exhibit and producing oscillators, switches, and antennas for stations. They additionally built four portable devices designed to be mounted on the backs of mules.[12] With an aim of modernizing further, the department sent prominent technicians to study abroad. José Flores Treviño and Raymundo Sardaneta studied Europe's most powerful stations, and

Pedro N. Cota went to New York to attend conferences held by the increasingly powerful Radio Corporation of America (RCA).[13]

Aficionados, Entrepreneurs, and the Rise of Commercial Radio

Thus far it appears that Mexican radio was largely a state project, if dependent on foreign manufacturers and training. But while the revolutionary government maintained a strong interest in radio, amateurs and local capitalists began to operate their own experimental stations. These enthusiasts mostly came from middle class and wealthy elements of society. They included military communication officials, sons of prominent capitalists, doctors, a communist leader, and engineering students. Most were influenced by the rise of similar experimenters and hobbyists in the United States and the American radio periodicals they followed. These amateurs and businessmen brought about a brief democratic opening in radio use and new challenges for the state, especially in matters of economic development, state propaganda, nation building, transnational messages, foreign relations, and antigovernment transmissions. Through legislation, international conventions, and less formal meetings, the private and government sectors ultimately compromised on these issues, reinforcing the consolidation of broadcasting into the hands of a few large corporations and the Sonoran leaders—Obregón, de la Huerta, and Calles—who had taken control of the nation's highest political offices.

Monterrey, a large industrial city close to the U.S. border, quickly became one of the largest epicenters for private radio use. Many privileged families in the region sent their children to the United States for higher education. These students became interested in U.S. technologies and trends. Born in 1898, Constantino de Tárnava Jr. provides the classic example. Born to Constantino de Tárnava de Llano and Octavia Garza Ayala, members of the Monterrey business elite, Tárnava Jr. came from a wealthy family.[14] Two years after his birth, his father became a deputy director and the first treasurer of the Compañia Fundidora de Fierro y Acero de Monterrey, a large smelting operation backed by local, French, and Spanish capital.[15] Tárnava Jr. first attended Saint Edwards College in Austin, Texas, and later Notre Dame University in Indiana. Engineering classes in America whetted Tárnava Jr.'s interest in broadcasting. Arguably, he, not the Gómez brothers, conducted the first ever broadcasts in Mexico as early as 1919.[16] On October 9, 1921, he began more regular broadcasts from his experimental station 24-A, Tárnava Notre Dame (TND)—named after his alma mater—which provided news and music to other fans, including Rodolfo de la Garza, the manager of the Bank of Nuevo León.[17] His station became CYO after receiving an official license and commercial status in 1923, and XEH in 1939. It remains the longest-running station in Mexico today.

Enthusiasts existed in other cities as well. Manuel Zepeda Castillo broadcast from the Teatro Degollado in Guadalajara and Tiburcio Ponce installed station 7-A Experimental in Morelia.[18] In Mérida, Arsenio Carrillo, the secretary of the Yucatecan Association of Radio-telegraphers, operated an experimental station and relayed programs from radio station PWX in Cuba.[19] By 1922 numerous other cities had broadcasting fans and participants, including Campeche, Xalapa, Oaxaca, Tampico, Veracruz, San Luis Potosí, Chihuahua City, Ciudad Juárez, Pachuca, and Cuernavaca.[20]

Like Monterrey, Mexico City was another center of radio experiment-ers. In the capital, however, they came from a slightly wider swath of society. Engineering students such as Enrique Vaca and José Peredo built their own radios from parts of other machines and instructions from U.S. magazines. Starting in the late 1910s, they listened to boat signals, radiotelegraphers, other radio aficionados, and after 1920, U.S. broadcasting stations. Adolfo Gómez, the man who organized the centennial broadcast, was a military doctor. His brother, like his father, worked as a dentist. José Allen, one of the founders of Mexico's Com-munist Party, transmitted messages from the barrio of San Rafael in Mexico City. Scientists and hobbyists spent long nights constructing radio apparatuses and other novel machines, rarely making much money from their efforts. Engineers in the Federal District universities and their counterparts in the military experimented widely with electricity and radio.[21]

Mexico City enthusiasts, engineers, and businessmen also formed the most powerful radio interest group in the country: the Liga Central Mexicana de Radio, or the Central Mexican Radio League (LCMR). Founded in early 1923, it was the result of the merger of the first Mexican radio society, the National Radio League established on July 6, 1922, and two other Mexico City organizations, the Central Radio Club and the Center of Engineers.[22] The LCMR's leaders not only possessed technical know-how but also connections within the business community and upper echelons of government. Modesto C. Rolland, the LCMR president in 1923, had been a mathematics and engineering professor under both Díaz and Madero, and a propagandist and a communications official for Carranza during the Revolution.[23] José M. Velasco, another important member, owned a prominent radio store in Mexico City and operated some of the country's best receiving equipment, picking up stations as far away as Salt Lake City in 1922. LCMR engineer Salvador F. Domenzáin installed radio equipment for Secretary of Foreign Relations Alberto J. Pani.[24]

American influence was not limited to programs and magazines. U.S. manufacturers increased their exports of radio receivers to Mexico. Successful commercial broadcasting began in the United States in 1920, and shortly thereafter American radio companies began selling their products in Mexico though Mexican partners, especially by advertising U.S. programs.[25] Some U.S. stations even provided shows targeting Mexican audiences.[26]

U.S expansionism dramatically affected the rise of broadcasting in Mexico.[27] Immediately after World War I and the decline of European influence in Mexico and Central America, U.S. manufacturers flooded Latin American markets with advertisements and goods. Radio was an important component of this economic surge, both as a communication tool for American businesses and as a commodity. The U.S.

FIG. 12. "For the Rabid Aficionado . . . ," *Excélsior*, December 7, 1924.

navy—after failing to convince Congress to continue a navy-controlled radio monopoly created during the Great War—helped establish RCA in order to buy out American Marconi and gain control of wireless communications in the United States and abroad. By 1921, RCA surpassed previously prevailing German and British radio interests in Latin America.[28] A friend of modernity and capitalism who knew that Mexico lacked the industrial wherewithal to compete, Obregón allowed, even promoted, this American invasion of goods. But he also encouraged the development of a local broadcasting system to counter U.S. influence.

In response to these developments and domestic proposals to build a radio-chain monopoly, Obregón asked a number of leaders and engineers to investigate the history of Mexican radio, recent developments in the field, and ways of regulating the technology considering its growing diversification and use.[29] Beginning in October 1922, these radio specialists, including longtime consultants such as the German World War I–era émigré Gustavo Reuthe, provided their opinions and accounts. All of them saw the new trends in radiotelephony—early broadcasting—in the light of previous radio operations, building their narratives on the foundation of the country's past relationship with radiotelegraphy. However, as some specialists admitted, the technology had outpaced the existing laws as stated in the 1916 radio decree and Article 28 of the 1917 Constitution. Many aficionados, entrepreneurs, and a number of SCOP officials believed that broadcasting paved the road to a new future for radio, where transmission of concerts, news, and lectures reached mass audiences. This perception, together with the growing reality, did not fit well with laws that addressed wireless as a form of point-to-point communication used more or less as a postal service and military tool.

The vast majority of advisers recommended a cautious liberalization of radio use. One government engineer argued that aficionados and experimenters who had been operating outside the law had resolved a number of technical problems and were responsible for much of the progress in radio development. Therefore he contended that they deserved more freedom to operate legally. Of course, the government had limited options in doing otherwise. Halting private use would have proved immensely difficult since many enthusiasts had already began to transmit experimental radiocasts and because U.S. and Cuban stations already broadcast programs that reached Mexico. As another specialist pointed out, receivers were fairly easy to make, and a number of residents had already constructed them.[30]

Every report to the president argued for strict government vigilance of private radio. Obregón's experts proposed that commercial companies should be allowed to grow—if the government did not want to exercise its right to monopolize the medium—but that they should all be licensed, censored, and operated by people with sufficient technical expertise. SCOP advisers continued that inspectors from their department should regularly check all transmitting operations. One expert listed a long number of concerns about espionage, using radio against the interests of the administration, divulging military operations, and interfering with national radiotelegraph stations—accurately foreshadowing problems that arose shortly thereafter. Others discussed the very real chaos in the United States that resulted from ill-regulated radio transmissions coming from thousands of radio hobbyists. Another issue was foreign concessions, which no one favored. All the advice was couched in patriotic, scientific, and moralistic prose, emphasizing what was best for national interests and development, what was best for the state: there should be nothing that "attacks the moral and healthy customs" of the nation, but the government should assist experimenters who work with the government, do not interfere with "national stations," and are not "dangerous for the nation."[31]

Aficionados themselves became influential advisers to Obregón, especially members of the LCMR. These specialists had intimate knowledge of the new trends in broadcasting and in mid-1923 approached the government about authoring new broadcasting regulations. Obregón agreed, but reserved the right to make changes he saw fit. Some scholars have claimed that the LCMR solely represented wealthy commercial interests, but that is not completely true. Its members came from business, amateur, and government backgrounds.[32] They were a fairly diverse, if mostly middle-class, bunch of characters. From the onset of broadcasting in Mexico, these engineers, entrepreneurs, and government representatives worked together for profit but also for various ideals of nationalism.

All in all, the LCMR-written regulations drafted between March and September 1923 provided a compromise. They addressed all the groups involved in radiocasting, but did lean in favor of business. They divided the ethereal space between experimenters, state broadcasters, entrepreneurs, and "public-service" radiocasters. The LCMR deemed public-service broadcasts as those that provided educational and entertainment programs of "public interest," whereas amateur broadcasts were those dedicated to experimentation.[33] The guidelines also established rules for licensing and taxation. They allowed for private profiteering, but the SCOP had to approve all permits for which businesses had to pay an annual fee. The rules restricted the power of private experimental transmitters to between 100 and 2,000 watts and limited the accessible frequencies to between 350 to 550 kilocycles. The reason for these rules was to decrease interference over the airwaves and thus to ensure the clarity of transmissions of commercial stations and military stations.[34] During the negotiations, Rolland argued for limiting experimenters and public stations from broadcasting during certain peak hours of the evening, suggesting his pro-commercial leanings. Yet, the following year the LCMR campaigned for amateur rights during the National Radio Convention. Rolland also defended aficionados against what he saw as unfair fees, though he ultimately bowed to Obregón's taxation policies. The LCMR additionally recognized and sought out the government as the protector of Mexican broadcasting, promoting the state's supervision over radio.[35]

Most LCMR members stressed elitist visions of civilizing the masses. One member, Argentina-born F. C. Steffens, disdained popular music. He

FIG. 13. Raúl Azcárraga, *El Universal Ilustrado*, April 5, 1923. Courtesy of Biblioteca Nacional de México.

and a number of writers and radio enthusiasts pushed the government to take a greater role in using the medium as a civilizing and educational tool. Rolland and Manuel M. Stampa, a prominent experimenter and officer of the league, declared that the LCMR's purpose was to coordinate the propagation of radio, which "suddenly puts the men of remote villages in contact with the civilization of the most advanced centers of culture."[36] When Rolland protested the tax put on wireless equipment, he stressed the detriment to education and human progress, not to enterprise (though he surely saw it as a deterrent to business as well). Rolland argued that the fee impeded the spread of receivers, which college classrooms and households could use to tune into transmissions of important conferences and lectures.[37]

The close interaction between the various groups involved with broadcasting was on display during the first commercial broadcast by CYL, *El Universal Ilustrado*–La Casa del Radio on May 8, 1923. Manuel Maples Arce, an Estridentista (or Stridentist) poet, delivered the first words. He leaned into the microphone and spoke: "Stars launch their programs at nighttime, over silent cliffs . . ."[38] A wide assortment of people accompanied the tech-enthusiast poet, including businessmen, politicians, musicians, and starlets. Carlos Noriega Hope, editor of *El Universal Ilustrado*, helped organize the event. He and *El Universal* editor Miguel Lanz Duret also represented their respective papers, one half of the station's partnership. The other partners, Raúl and Luis Azcárraga, owners of radio store La Casa del Radio, socialized with the attendees and participants. The head of SCOP, and former leader of Venustiano Carranza's revolutionary military telegraphers, Amado Aguirre, joined the celebration as a sign of government support and supervision. Andrés Segovia, the increasingly famous Spanish guitarist, played his instrument to great applause. Mexican singers, musicians, and composers came to showcase their talents for the event as well. Actress Celia Montalván and vocalist Julia Wilson de Cháves sang various popular songs, and pianist Manuel Barajas and the renowned composer Manuel M. Ponce played original classical pieces. Before signing off and leaving

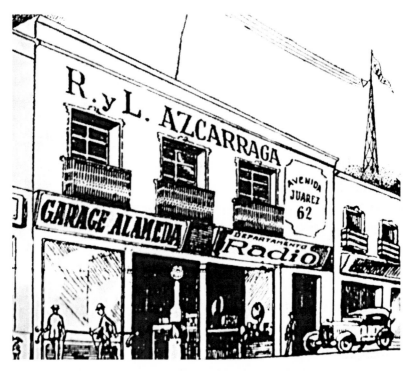

FIG. 14. Raúl and Luis Azcárraga's El Garage Alameda and La Casa del Radio, home of CYL, *El Universal*, May 20, 1923. Courtesy of Archivo General de la Nación de México.

the listeners to the static, the function concluded with a broadcast of the Mexican national hymn.[39]

This inaugural radiocast was a foretaste of the country's multicultural future, a soundscape combining the works of professionally trained composers, folk songs, and modern musical trends from the United States, Europe, and other Latin American countries. The program included avant-garde artists and editors who promoted a metropolitan and global worldview, nationalist composers who disdained popular shows, popular performers, a foreigner with monarchist sympathies (Segovia), and a revolutionary communications official. But as Barajas would make clear the following year, not everyone agreed with this potpourri approach to radio.[40] Indeed, broadcasting unleashed a cultural debate about the medium, music, and Mexican nationalism.

The first commercial broadcast also exhibited the partnership between the new state and rising business elites. The Azcárraga family had obtained a small fortune selling Ford cars in Monterrey during the late Porfiriato and during the Revolution. Emilio Azcárraga, Raúl and Luis's brother, had just recently married the daughter of Patricio Milmo, one of Monterrey's wealthier men and a member of the Monterrey Group of industrialists. The Obregón government backed the Azcárragas, participating in their opening broadcast and providing public praise. The partnership not only allowed private enterprises to shoulder most of the cost of broadcasting, but also helped form a bridge between the revolutionaries in power and businessmen who had not originally supported the Revolution.[41] Indeed, the Azcárraga operation became one of the biggest outlets for state leaders and, in turn, usually operated with government approval. The relationship between the Azcárraga family and the government would remain strong, though often privately contentious over issues of state broadcasting, religion, and politics.

After the premiere CYL airings and the creation of the first broadcasting regulations, the LCMR, along with government officials and a multitude of businesses, organized the Grand Radio Fair in Mexico City. It was an impressive ten-day exhibit of radio broadcasting and products. Lasting from June 16 to 25, booths displayed locally made radios and the latest receivers from the United States. Participants visited the elaborate stand, admiring the goods, contests, and wacky costumes. In one of the most outlandish displays, women gave away El Buen Tono's "Radio" cigarettes while wearing mock antenna hats that looked like diamond-shaped kites. The tobacco company was in the process of constructing the country's second commercial station, over which it would continue to promote its cigarettes. In competition with El Buen Tono, CYL representatives handed out "Radio" sodas. Obregón inaugurated the event, posing for photos ops with Rolland and the various businesses. Government radio stations JH–Secretary of War and Marine and VPD–Department of Military Manufacturing both aired music, including the first broadcasts of Mexican military bands.[42]

FIG. 15. Ad for El Buen Tono's Radio cigarettes by Fermín Revueltas, *Irradiador* 1, no. 3 (November 1923). Courtesy of the Hamilton Library, University of Hawaii.

The collaboration between the government and commercial broadcasting companies is seen in another important way as well; many of the station builders and operators worked simultaneously for the most important state and commercial stations during the 1920s. The CYB–El Buen Tono station provides a perfect example. Inaugurated in September 1923, CYB became one of the most popular stations in Mexico during the 1920s. El Buen Tono, founded by Ernest Pugibet during the Porfiriato, remained one of the most profitable enterprises in the country, surviving and even profiting during the chaos of the Revolution.[43] Although Obregón pressed for unpopular taxes on industrialists and catered more to worker demands than previous presidents, he generally supported Pugibet's successors at El Buen Tono along with many other Porfirista entrepreneurs. The company's leadership, on the other hand, co-opted many revolutionary technocrats by giving them positions. Many of the most important members of CYB worked, or had worked, for the revolutionary state. Colonel José Fernando Ramírez, instrumental in the creation of army station JH, worked for CYB and other commercial stations. So too did former JH designers and operators Captain Guillermo Garza Ramos and José de la Herrán.[44] José Reynoso, manager of the tobacco factory and the station, was a senator of the state of México from 1917 to 1920.

Obregón himself was an "ardent radio fan."[45] He kept in direct contact with the specialists developing and regulating the medium. He not only inaugurated the 1923 radio fair, but also lent his voice to radio stations. CYR–Roserter y Cía in Mazatlán, for example, commenced in the autumn with a message from the president.[46]

Before Obregón left office, broadcasting stations had expanded significantly. The "Hour of Transmissions" section of *Excélsior* in October and November 1924 reported five aficionado stations, one pro-Calles labor party station, and six commercial stations. The central government operated three stations. The Chihuahua State Telephone Department ran another. Dozens of government, commercial, and amateur radio-telegraph operations existed as well. At the same time, Mexican radio

FIG. 16. El Buen Tono ad, *El Demócrata*, August 26, 1923. Courtesy of the Archivo General de la Nación de México.

listeners could pick up over a hundred U.S. broadcasters, a continual presence in Mexico's electronic soundscape.[47]

Radio receivers had gained popularity in urban Mexico rather quickly. Following the Grand Radio Fair, the U.S. Department of Commerce stated that radios were in high demand in Mexico. Journalists reported that show windows were filled with radios, often blaring Mexican or American radio stations to the amusement of large crowds. Although most popular in Mexico City, radio had dedicated enthusiasts and salesman in Monterrey, Tampico, Mérida, Oaxaca, Mazatlán, Chihuahua, and other regional centers.[48]

Foreign Relations

In addition to supporting and regulating commercial broadcasting, the Obregón administration continued to play a vital and direct role in radio in other ways, especially in foreign relations. Its station in Chapultepec relayed news from abroad. SCOP director and future president Pascual Ortiz Rubio ordered the station to provide the Mexican press with the foreign news it received.[49] Mexico continued to build radio towers in Central America. The state also played a crucial role in protecting the fledgling commercial industry at an important international radio conference.

From his first days in office, Obregón built on the initiatives of his predecessors to enhance foreign relations through radio. This effort was crucial for Obregón's drive to gain support from foreign leaders from 1920 to 1923, when the U.S. government withheld recognition. As a result, Obregón tried to influence policies via radio-listening audiences in the United States and wireless communications with German, Japanese, Salvadoran, and South American officials.[50] For their part, most Latin American governments recognized Obregón's administration and chastised the United States for not doing the same. This stance was made clear when their representatives loudly vocalized complaints about nonrecognition at the Pan-American Conference held in Santiago, Chile, in March 1923. Mexico did not attend in protest of U.S. polices.[51] The convention did, however, discuss inter-American communications, especially radio. The participants decided to hold a separate pan-American gathering specifically on electronic communications the following year, in Mexico City.

Obregón's also expanded the Central America radio endeavors of his predecessors. Building on Carranza's work in El Salvador, Obregón promised at the 1921 centennial fair to develop stations in Central America for a number of countries, specifically in Costa Rica, Guatemala, Honduras, Nicaragua, and to improve the Salvadoran operation. The target date for completion was Mexican Independence Day, September 16, 1923. The secretary of foreign relations, secretary of communications and public works, and the director of the DGTN all backed the plan. For the equipment, the government turned to their oldest ally in radio communications, Telefunken.

Some ambassadors in Central America, however, had serious doubts about the project, especially the ministers to Costa Rica and Nicaragua.

The latter warned that the "unscrupulous" governments in the region might use the machines against the interest of Mexico, a curious statement since he continued in the next sentence to say that he had little faith in the ability of Central Americans to operate and maintain the devices, arguing that they did not have the money or talent. He declared that in the near future the radio towers would be "outdated and useless . . . a mountain of old iron" gifted by the Mexicans.[52] Eduardo Ortiz, Obregón's nephew and the consul in Costa Rica, also vocalized fears about possible U.S. reactions to the devices and complained about the costs to the federal treasury.[53] He provided a counter idea, to forgo the stations and to instead build "Mexican" libraries for the isthmian workers. Ortiz even went as far as to draw out elaborate blueprints for his imagined facilities. This concept fit into Ortiz's idea that the Mexican government should directly address the people of Central America more than government officials.[54] However, Obregón had already committed to the radio idea.

There was less reserve from Guatemalan representatives. Baltazar Chávez, a Guatemalan official who resided near where the radio station was being built, wrote directly to Obregón, giving the sincerest thanks from the city of Quetzaltenango and the whole of the country for gifting the wireless station that was "attended by personnel of all social classes, having great animation and affection for Mexico."[55] The consul to the region, Juan de Dios Bojórquez, consistently nudged Obregón to get the needed materials to Guatemala to speed along the production.[56] In response, the president directly inquired into the SCOP's progress on the Central American projects, urging them to complete the task.

The assignment was not an easy one. There were a number of complicated logistics involved. Much of the equipment had to make its way across the Atlantic from Germany to Central America, where it then had to be carried overland through rainforests and across undeveloped terrain. Most problematic, the government struggled to cover the costs, lagging in its payment to Telefunken. On June 13, 1923, SCOP employees crossed the border into Guatemala to ready the site for the construction of the station's buildings. But back in Mexico City their fellow workers were still fabricating the necessary iron towers in the workshops in Chapultepec. Meanwhile, the actual receivers and transmitters had not yet left Germany because the Obregón administration still owed Telefunken the equivalent of 25,000 U.S. dollars.[57] The government also possessed a limited number of radio specialists such as Luis Sánchez and Gustavo Reuthe who were capable of building these operations, which had to have put a constraint on production.

Despite the difficulties and over the protests of some of his ambassadors, Obregón pushed ahead with his wireless plans in Central America. And though not all of the operations were in full working order by September 16, 1923, they were at least under construction. Realizing that Obregón would not change his mind, Ortiz changed his tone. He posed for photo ops with the Costa Rican president Julio Acosta García at the commencement of the station's construction. Their wives christened the radio tower with a bottle of champagne during the Independence Day festivities.[58]

The program of donating radio stations to Central America fits interestingly into Obregón's overall foreign policy in the region. Obregón faced a number of tricky problems in hemispheric relations. For the first years of his presidency, he needed to be careful not to raise the ire of the U.S. government if he wanted to obtain recognition, which would aid him economically, politically, and militarily. At the same time, a large component of his own followers pushed for a more radical, nationalist, and at times, anti-American stance, including many of the ambassadors to Central America. They believed that they needed to strengthen Mexico's presence and influence in the isthmus to build up Mexico's prestige and to deter America's imperialistic tendencies.

Obregón approached the area with caution, but his policies were perhaps not as "hands off" as some scholars have argued.[59] To be certain, he did not meddle heavily in the military and political affairs of the region, keeping a low profile, but he did more than lay foundations for the future. He and his ambassadors established a greater cultural presence, including a number of libraries that the administration decided to build in addition to the radio stations. And although the radiotelegraphy towers were not set up to broadcast, the Costa Rican government, at least, worked to acquire the equipment to adapt the station to both forms of radio.[60]

Radio receivers in general were on the rise in Central American cities. By the end of 1923, Central American urbanites heard a number of Mexican broadcasting stations including CYL, CYB, CYX–*Excélsior*–La Casa Parker, and CYF, in Oaxaca City. Before the end of the following year, they could pick up the government's own public education operation, CYE, soon changed to CZE.[61] Multiple stations joined them in international broadcasts over the course of the decade, laying the ground work for a strong Mexican influence in the region's electronic media. These stations aired Mexican nationalist composers, foreign and domestic popular music, presidential speeches, hygiene lectures from the Department of Public Health, sporting events, news, and advertisements for Mexican beer and tobacco products.[62]

Underneath this cultural campaign, the stations gifted to Central America helped Obregón achieve other goals. For one, he acquired more efficient communications with the region's governments while building a stronger coalition against American cultural expansionism. It also allowed his administration to work against U.S. attempts to dominate communications development in the isthmus. With the blessing of their government, American agricultural outfits, especially United Fruit, had largely controlled wireless operations in Central America until the 1920s, owning the only stations in addition to their own small fleet of ships equipped with wireless devices.[63] Obregón's radio campaign impeded American communications dominance by building a Latin American coalition to increase state oversight of radio development and to oppose American initiatives for a free-market policy. The Costa Rican press exhibited this statist position following the groundbreaking for the radio tower donated by the Mexican government, stating—contrary to American interests—that Costa Rica would build a nationalist wireless system, structured on a government monopoly of radio services.[64]

This position for Latin American unity against the United States on the matter of wireless communications was in full display at the Inter-American Committee on Electronic Communications held in Mexico City in 1924. Convention participants stated publicly that the goal of the meeting was to improve Pan-American telegraph and broadcasting operations, but another consideration soon became apparent as the delegates of the fourteen countries divided into two camps. On one side, the United States promoted private capital-driven communications development with limited government intervention. On the other side stood the Latin American representatives who pushed for a more state-directed policy toward radio. Ignoring the subtleties of the issue, U.S. spokesmen claimed that the argument revolved around commercial versus state-run radio.[65] But the Latin American stance against U.S. policy did not mean that they opposed the development of private communications industries within their perspective countries. Quite to the contrary, some countries, Mexico included, pushed for greater state regulation of wireless telegraphy and radiocasting to protect the growth of newly initiated commercial broadcasting industries. For Mexican officials, it was also about sovereignty. Radio was one of the few "public works" that foreigners never controlled. The problem was that Latin American governments feared the domination of American businesses at the expense of native companies and governments.

All the Latin American delegates in attendance voted against the position of the United States and accepted the guidelines put in place by the Mexico City conference. The American representatives abstained from voting. However, the rules established at the 1924 meeting had little lasting impact on international communications policies. Only four of the attending nations ever ratified the agreement.[66] Instead, they decided to wait until the next meeting, which was announced in 1925 and took place in Washington DC, two years later. But the 1924 conference did exhibit the Mexican government's long-held unwillingness to allow U.S. corporations to control Mexican radio. The meeting also helped create a framework that the legislature used for federal radio legislation, especially the 1926 Law of Electronic Communications, which provided the government more regulatory power over broadcasting.[67]

The Lingering Specter of Civil War: The De la Huerta Revolt

In December 1923, Adolfo de la Huerta violently rebelled against the government he had helped establish, provoking officials who remained loyal to the state to reexamine their views on radio. The revolt caused lasting ramifications for wireless development in Mexico, especially increasing authoritarian and monopolistic tendencies. The rebellion was a mix of genuine concerns about the future of democracy in Mexico and greedy ambitions for power. Calles had obtained the official support of Obregón, making him the favorite to become the next president. Feeling disregarded and upset at the unfair influence of the president, de la Huerta obtained the support of military leaders disgruntled about Obregón's reduction of the armed forces or possessing their own ambitions for power. On December 7, they rebelled against the "impositionists."[68]

Radio proved crucial to the rebellion and its suppression. By late 1923, wireless technology's importance as a tool of espionage and military communications had grown significantly. Radio was not auxiliary, but essential. It had become a much greater force for distributing messages and spreading propaganda. There were now thousands—an estimated fifteen thousand in Mexico City alone—of receivers dispersed across the country and even more in the neighboring United States.[69] Radio allowed de la Huerta to spread word of his cause and to counter government stories printed in the press. In turn, jamming and intercepting de la Huerta's wireless correspondence was a constant goal of government forces, which then spread their own radio propaganda on an even larger scale. War had become internationally audible to a much greater extent.

The De la Huerta Rebellion posed a serious threat to the Obregón government, quickly becoming a full-scale civil war. Nearly half of the military rose against the president. Rebels controlled many of the important ports and radiotelegraph stations, and the initial drive on the capital caught the government off guard. In western Mexico, General Enrique Estrada, along with a number of other high-ranking military officers, four regiments, and three battalions, mutinied, seizing control of Guadalajara and much of the Jalisco countryside. In the state of Tabasco, General Ferrara Vega joined the rebellion along with much of the army under his command. General Fortunato Maycotte, after persuading Obregón to give him 200,000 pesos and war supplies to suppress the rebellion, immediately switched sides, rallying insurrectionists in Oaxaca.[70] In the east, Delahuertistas captured the state capitals of Xalapa, Veracruz, and Puebla, quickly moving nearer to Mexico City. Many residents believed that Obregón was going to abandon the capital in order to fight from another location.[71] Momentum was on the rebel side. The revolt, however, stalled after distrust and disagreements within the leadership halted the advance on Mexico City, which allowed Obregón to launch a successful counterattack.[72]

Radio espionage provoked a strong fear and response in the Obregón administration. Immediately following the declaration of rebellion, the government ordered the suspension of all internal wireless communication with rebel-controlled areas, including Túxpan, Veracruz, and Campeche.[73] In order to uncover insurrectionary operatives in Mexico City, the military and SCOP used special agents to search the urban airwaves for renegade messages and to locate their bases of operation. José Soto worked as one of these spies. One of his missions included searching for radio listeners and transmitters in the affluent neighborhood of Santa María la Ribera. In mid-January 1924, he listed thirty-nine households with receptors in an eight-street area.[74] Soto worked hardest at discovering the meeting place of a group of subversives who aided the Delahuertistas cause.[75] His superiors, including the commander of the Federal District, General Arnulfo Gómez, specifically targeted Jorge Carregha, a known conspirator whose house had a radio station with a transmitter and receptor for communicating with the rebels.[76] Soto also vigilantly watched the house of Ignacio Flores, a suspect connected to Carregha who lived on Flores Street and owned a "very well mounted" transmitter.[77] The threat of radio spies was real. They existed.

To compliment Soto's work, the government put in place new restrictions on radio use in the capital. This led the paper *El Universal Gráfico* to publish a headline "Formal Raid against Radio Apparatuses."[78] After intercepting a number of insurrectionary messages transmitted from the capital, Gómez restricted all transmissions to licensed commercial stations and those under direct approval and surveillance of the armed forces. Aficionado radio stations had to stop their operations. Others had already been shut down by state officials. The military further ordered all residents with radio receivers to register their names and equipment with the government. Those that failed to do so were condemned as spies and enemies of the state. Radio providers had to inform the garrison commander of any sales, providing the required information before the transaction could be completed. The government outright forbid the sale of radio transmitters.[79]

The administration was not merely paranoiac. Rebels did rely heavily on radio to communicate with spies and different regional commanders. The revolt was widespread. Radio became a regular means of communication between leaders in Veracruz (and then later in Frontera, Tabasco) and coconspirators in Jalisco, Oaxaca, and the Yucatán peninsula. Estrada not only communicated his entrance into the rebellion and the strength of his forces via radio, he also transmitted regularly to other generals, including Maycotte in Oaxaca via the Salina Cruz station. In late December 1923, rebel general Rómulo Figueroa used radiotelegraphs to discuss the movement of his forces from the state of Guerrero to Morelos. In Mérida, Yucatán, rebel forces under Ricárdez Broca took possession of a number of radio sets used by aficionados. The insurrectionists also took over the government radiotelegraph station, murdering Remigio Ortegón, the station chief.[80] Controlling radio communications was an early goal of the various Delahuertista forces and crucial to de la Huerta's ability to keep in contact with the rebellion's leaders in other parts of the country.[81]

But more often than not, these wireless links so crucial to Delahuertista operations worked against them. Government officials regularly intercepted messages that provided much of the intelligence on rebel forces and movements. In mid-January, as a significant battle over the city of Puebla neared, military radio specialists overheard a conversation between Generals Maycotte and Estrada about Delahuertista forces abandoning the area around Tehuacán, Puebla, and moving in the direction of Esperanza to the north. After the fall of Puebla, during the march on Veracruz, military radio bulletins stated that "wireless messages exchanged between [Gen. Guadalupe] Sánchez from Veracruz and various rebel commanders, which were intercepted at the Chapultepec station, indicated that Adolfo de la Huerta was making a desperate effort to recapture San Marcos [Veracruz]."[82] Government stations also picked up on the widespread disagreements between Jorge Prieto Laurens and other important rebel leaders, and that de la Huerta had requested a hundred men from Villahermosa to act as his personal escort. In response to the news that Figueroa had moved into Morelos, the Department of War sent General Gómez and a column of "1,500 infantry, cavalry, artillery, and two airplanes" to reverse the rebel advance."[83] Similar radio interceptions and subsequent responses continued throughout the conflict.[84]

In February and March 1924, Otilio González stepped up the rebel wireless propaganda campaign as their military situation worsened. From his newly established headquarters in Frontera, Tabasco, de la Huerta sent out a radio "Manifesto to the Nation" on February 20. From New Orleans, Arturo M. Elías relayed the speech in its entirety to his half-brother, Secretary of War and presidential candidate Plutarco Elías Calles. Elías obtained the information from the crew of the warship *Bravo*, who overheard the speech while the vessel was being refitted in New Orleans. The radio address touched on all the grievances set out in his original declaration of rebellion before arguing that the Obregón government had violently expanded the war across the nation and had sold out to Americans in exchange "for ships of war, airplanes, rifles, projectiles, and money."[85] González and rebel radio operators in Mérida sent regular messages to the Associated Press in Dallas, Texas, which newsmen subsequently spread to America's largest papers, including the *New York Times*. González strove ardently to paint a picture of rebel victories, with some success, even in the face of a horribly deteriorating situation from late February through April.[86]

The worsening situation inspired de la Huerta to flee Mexico on March 11. Departing in a small boat, he stopped briefly in Carmen, Campeche, where he met Prieto Laurens who gave him 14,000 silver pesos. Telling Prieto Laurens that he was heading to Campeche and then to Yucatán, de la Huerta actually boarded the steamer *Tabasco*. On board he bought the loyalty of the crew with his recently acquired pesos, which in turn, refused to answer the calls from the sailors of the modified corvette *Zaragoza*, whose loyalty to the rebellion had become suspect.[87]

Of course, the Obregón administration had its own radio operations. Although the vast majority of the Gulf Coast fleet joined the rebellion, the four cruisers obtained from the United States were equipped with radio stations. As previously mentioned, Americans also refurbished the gunship *Bravo*. These ships helped secure the government's victory in Tabasco, Campeche, and Yucatán. Airplane squadron commander Captain Rafael Ponce de León used a radio device in his plane to relay information on the location, numbers, and movement of enemy forces—information that was immensely valuable. SCOP officials had recently installed an impressive radio car in the presidential train, an accommodation that Obregón converted into his moving military quarters during the war. The president's own personal telegraph corps operated the wireless equipment, including DGTN employees Manuel Serrano and Federico W. Kreush y Arce.[88]

Colonel José Fernando Ramírez and Captain Guillermo Garza Ramos, who both helped build JH, CYL, CYB, and radio station *El Mundo*, constructed receiving towers "in all the military camps" to pick up long-distance wireless messages, speeding up communications from Obregón and other top generals to the field and allowing for greater coherence during mobilizations. These devices also intercepted enemy messages. Ramírez and Garza Ramos set up the equipment and also trained field operators on how to use it.[89]

The memoir of General Donato Bravo Izquierdo provides a number of insights in how government forces incorporated radio in Chiapas. Using a radio set up by Lieutenant Pedro Ríos in a small brewery, the general was able to hear messages about the war front in Veracruz, including the defeat of Guadalupe Sánchez in Esperanza. Using the recently donated stations in Guatemala and El Salvador, Bravo Izquierdo relayed messages to Mexico City and receivers in Veracruz. These countries also provided ammunition to his forces.[90]

Military and political leaders also used radiotelegraphs in more personal ways. Aboard President Obregón's recently revamped presidential train in Guanajuato, political ally Fernando Torreblanca sent wireless messages to his wife in Mexico City, giving "mucho, mucho, mucho, cariñosos recuerdos," and making sure that his son brushed his teeth.[91] Keeping in close contact with loved ones was a perk of being close to top-ranking officials and their affiliated radios.

In addition to radiotelegraphy, the SGM aired its own propaganda via broadcasting, highlighting its victories but also mentioning some of its setbacks.[92] From Mexico City, the department sent out bulletins on its own stations, as well as on all other government stations. The Chapultepec transmitter sent these reports across the hemisphere and to western Europe. The receivers among the Regional Confederation of Mexican Workers (CROM) and affiliated factories spread government propaganda among urban laborers.[93]

The commercial stations also became embroiled in the conflict. Immediately before the official outbreak of the rebellion, Martín Luis Guzmán, a de la Huerta supporter who had planned to use his station *El Mundo* in support of de la Huerta's campaign, abandoned the operation, and his newspaper of the same name, for exile abroad.[94] The government also briefly shut down CYL. Shortly thereafter it reopened, airing daily war bulletins provided by the SGM.[95] CYB continued to broadcast performances by the National Conservatory of Music and Theater, apparently posing no threat.[96] The rebellion sent a clear message to radio operators: if you go against the government you will be shut down, but if you support the government, it will support you. In Mexico City, questions of loyalty had largely been determined before the end of January. Once the Obregón administration decided that a station would help and not hinder its military operations, officials allowed, even encouraged, commercial broadcasts as a means of building support and further advancing the private broadcasting initiatives began the year before. For almost the entire conflict, commercial stations continued to operate.

Returning to the New Normal

By late April, a month after de la Huerta fled, the revolt had mostly subsided and radio development continued much as it had before. Intercepted wireless messages showed the remaining rebels lacking any cohesion and evacuating their remaining strongholds in the south. The landing of government forces in the Yucatán peninsula on April 21 put an end to the last serious bastion of rebel support.[97] Although pockets of insurrectionists continued to operate for a number of months, they received less attention in Mexico City and abroad. The rebels had done significant damage to important radio stations, including in Veracruz, Mérida, and Salina Cruz, but government workers were already in the process of repairing them.[98] During the preceding month, the LCMR began to redouble its efforts to expand radio construction and use, providing a series of lectures, broadcasts, and conferences.[99] Commercial stations once again expanded in number during the last half of 1924. The Inter-American Committee on Electronic Communications, held in Mexico City in June, reestablished Mexico as a leading Latin American country in communications development. However, the rebellion had instigated an important new change: political speech became solidly monopolized by a single political faction and commercial stations toned down any dissidence.

While the military was delivering its final blows to the rebels, Calles and his supporters returned to the campaign trail, giving the first radio address by a presidential candidate on April 12. Showing their continued cooperation with the state, the Azcárraga brothers, via CYL, transmitted the program. The Progressive Civic Party (PCP), "the Party of the Middle Class," planned the event. Architect Guillermo Zárraga, the party's president, worked with the Azcárraga brothers to create the program. It consisted of the national anthem, military bands, classical music—both foreign and nationalist—and Mexican popular songs, squeezed in between the speeches of Calles and two leaders of the PCP.[100]

Interestingly and importantly, Calles and the PCP aimed their speeches at the United States as much as Mexico, "to the entire country and to the foreign countries where this powerful station reaches."[101] They addressed U.S. fears that Calles was "absurdly radical." The PCP justified Calles's proposed policies by connecting them to the ideas of social justice that U.S. president Woodrow Wilson espoused following World War I. They also argued that Calles would expand education and incorporate rural campesinos, urban workers, and the middle class, exhibiting the government's escalation of populist politics. The PCP also painted Calles as countering the previous wrongs of the landed aristocracy and the Catholic Church (though they did not address the Church by name). However, the PCP attempted to make these plans more acceptable to U.S. officials by emphasizing that Calles would rule democratically and by the law, that he was practical, compassionate, "more human and more scientific." Following the speeches delivered in Spanish, a translator provided them in English.[102]

The radio address was one component of a larger campaign that Calles carried out to ease the fears of Americans who saw Calles as representing the "Danger of Bolshevism under the Coming Regime."[103] U.S. officials had grown impatient with talks of revolution. Calles, like Obregón, walked a tightrope, arguing that he supported the Revolution, workers, and Mexican autonomy, but also that he was not the enemy of foreigners, foreign capital, and the sacredness of contract.[104]

Although CYL did reach portions of the United States, the American press and government appear to have had little reaction to the Calles-PCP radio event. The *New York Times* never discussed the address; instead, the editors gave more attention to Calles's speech to farmers in Morelos the following day, where the candidate paid lip service to the legacy of Emiliano Zapata, giving the paper another opportunity to paint Calles as a radical.[105] Calles's overall campaign to appear more moderate to Americans did, however, have some effect when during the following month other newspapers began publishing articles on his increased moderation toward American capitalists.[106]

It is hard to judge how much affect the speeches had on the Mexican public, but they do appear to have had some influence. But just in case Mexico City residents missed the address, *El Demócrata* dedicated a total of five pages over a two-day period discussing the event. The paper called the radio address "a great success" and said that it was heard clearly throughout the capital.[107] Following the event, members of the agrarian group Partido Rojo del Sur Veracruzano—which had fought against the De la Huerta Rebellion in Veracruz—sent the presidential aspirant a letter praising his defense of campesino rights, offering their support to his campaign, and stating that they heard his address clearly on their radio.[108]

The Partido Rojo del Sur Veracruzano radio and response exhibit another important trend. In addition to more authoritarian tendencies, the most important change to occur in state radio practices during and after the De la Huerta Rebellion was the government's push to use broadcasting and the gifting of radio receivers in a populist attempt to build a larger coalition of support, a point discussed in greater detail in chapter 6. Government leaders increasingly used radio to directly communicate with the people of Mexico as a means to solidify their power. The rebellion reminded Obregón, Calles, and their supporters that their grasp was fragile and that the country remained divided. Although the practice began before the rebellion, state officials showed a new impetus toward expanding direct radio communication with those that supported them, the rest of the nation, and the world abroad.

Conclusion: Crucial Years

The years during Obregón's presidency proved crucial to the intertwined developments of technological growth and corporatist organization in the revolutionary government. Airplanes were becoming a common component of the military. Engineers and construction workers erected large buildings of grey reinforced concrete. Camaras, cars, and typewriters continued to find their ways into the hands of people in towns across the republic. Articles in newspapers and magazines argued about the merits and pitfalls of technology, while a new group of radical poets—the Estridentistas—praised the new, modern, and metropolitan world that Mexico was entering. The Obregón administration wrestled with these fast-paced changes while struggling to maintain and expand its power.

The evolution of radio is a good example of the significant changes under way. During the early 1920s, the military equipped airplanes with wireless devices. Scientists and engineers used the technology to standardize time. From Chapultepec, government officials expanded their communication with the United States, South America, Japan, and Europe. Hobbyists searched the airwaves for distance messages and held competitions for the best homemade radio equipment.

The most significant change, of course, was the initiation of broadcasting. Radio amateurs, many of whom had been operating illegally, expanded in number and became some of the most significant promoters of the medium. Indeed, they were responsible for organizing many of the earliest broadcasting events and for the creation of the first broadcasting regulations. The military commenced its own broadcasting station, and so too did the SEP. It was also during the Obregón administration that commercial radio became a facet of society, especially in Mexico City, Monterrey, Guadalajara, and other urban areas. Throngs of people crowded around shops blaring programs from stations such as CYL and CYB.

Obregón's decision to let commercial enterprises take the lead in radio broadcasting signaled a new opening in radio use, recognizing the limits of the government to develop an adequate broadcasting chain, but also the capitalist leanings of the Sonoran leadership. However, the government did not give these stations free reign. The 1924 election and the resulting De la Huerta Rebellion made the administration much more hesitant about what private stations aired. Military officials closed down stations, permanently dismantling the one station that supported de la Huerta. During the uprising, the government demanded that all radio owners register their devices. Surely not everyone abided by these orders, but many did, and the decree shows the willingness of the Obregón administration to hinder democratic trends in radio development in the name of state control. The rebellion also had other significant ramifications for wireless development. It ultimately led to the increased dominance of armed forces loyal to the state in the military use of radio. Most importantly, it further cemented the relationship between the Azcárraga family, their station—CYL—and the revolutionary state. This bond is exemplified not only by the station's agreement to broadcast pro-government messages during the De la Huerta Rebellion but also by CYL's hosting of Calles's PCP event—the first-ever broadcast in Mexico by a presidential candidate.

5 Invisible Hands

> Man may make a perfect machine, but it still will depend on man himself as to whether the machine shall be an instrument of understanding or misunderstanding.
>
> Dwight Morrow, 1930

During the years of Calles's greatest political influence, from 1925 to 1935, radio and the revolutionary state matured together. The biggest change in wireless trends was undoubtedly the expanding broadcasting operations, but broadcasting was only one of the fields of radio that interested political leaders. The central government continued to invest in radiotelephony and radiotelegraphy for military use, foreign relations, and internal communications more generally. During this period, military commanders began a more genuine attempt to professionalize the armed forces, which included greater incorporation of radio. Private enterprises also became involved in wireless communications outside of broadcasting. Radio services increased in the transportation sector. All of these advancements, in turn, influenced the government to create more detailed legislation to regulate radio use. Although most aspects of radio before the Lázaro Cárdenas presidency have been underexplored, these components of wireless development have received even less attention, despite playing a significant part in the creation of modern Mexico. Military and SCOP radio projects were vital in securing the Calles government and the subsequent administrations during the Maximato and Cárdenas years.

Military Developments

Under the leadership of Minister Joaquín Amaro, the SGM advanced radio in the armed forces while managing to temporarily decrease overall costs.[1] The military not only continued to incorporate the technology for combat and administrative communications but also, along with cinema and print media, for "culturization" projects.[2] Radio was an important aspect of army reforms and was included in every branch of the military. Ground and air forces used radio equipment in all of the major campaigns of the era, including the Yaqui Uprising (1926–27), the Cristero Rebellion (1926–29), and the political revolts of Arnulfo Gómez and Francisco Serrano in 1927 and José Gonzalo Escobar in 1929. Unlike the preceding De la Huerta Rebellion, these insurrections, though serious, failed to gain the popular support of the military, a testament to the success of weeding out discontented officials and the formation of a more loyal army under Amaro's tenure. Greater cohesion within the military also allowed the state to control radio communications to a greater extent, lessening the danger of the medium being used for purposes contrary to those of the government.

From 1925 to 1933, the military's radio operations expanded significantly under Amaro and the brief directorships of Generals Abelardo Rodríguez and Pablo Quiroga. By 1931 the military had succeeded in placing radios in all of its twenty-four *jefaturas*, or military headquarters.[3] The SGM also began developing new "backpack" radios, which weighed 28.5 pounds, in addition to importing similar equipment from Britain and France.[4] Plans were under way to provide radio devices to all military units, similar to practices in the United States and Europe. Built in army workshops, the SGM produced devices for infantry battalions and Mexican cavalry regiments.[5] The rate of radio production substantially outpaced what had been built in previous years.

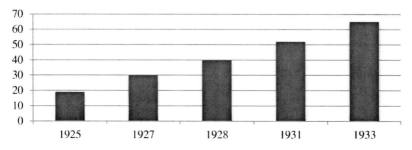

FIG. 17. SGM radio stations, including portable sets, 1925–33. According to one newspaper article, the SGM workshops were supposed to produce over a hundred field radios for different military units in 1933 and 1934. *Sources*: SGM *Memorias*, 1924–33; Estado Mayor Presidencial, "Informe de la Comisión Intersecretarial de Radio," May 1933, APECFT, Archivo Plutarco Elías Calles, exp. 95, inv. 1926, leg. 1; "Todos los cuerpos del ejército van a tener estaciones de radio," *Excélsior*, February 22, 1931, APECFT, Fondo Joaquín Amaro, exp. 255, inv. 566, leg. 1.

To work the new equipment, the armed forces trained more operators. On New Year's Day 1930, Amaro inaugurated the School of Military Transmissions, which had existed in a more experimental form as the Military Telegraphic School. The next year the school had over sixty soldiers studying military communications. In 1932, during the directorship of Abelardo Rodríguez, officials once again changed the school's name, this time to the School of Connections and Transmissions. That year the SGM possessed eighty-five officers and thirty-one soldiers specifically designated for military transmissions. These radio specialists increasingly connected army bases and incorporated new equipment. Every major military outpost possessed at least two radio operators.[6]

In the navy, meanwhile, officials worked to rebuild their small flotillas following the mass defection during the De la Huerta Rebellion. In the process, they upgraded or installed long- and short-wave radios on a number of their vessels. During the summer of 1927, the U.S. company Johnson Iron Works installed the most advanced RCA long- and short-wave radio equipment on the cruiser *Anáhuac*, which the navy had bought in the United States during the De la Huerta Rebellion.

FIG. 18. Military communications cadets in the School of Transmissions, 1929. *Memoria por el SCOP*, 1929–30.

Other ships with wireless devices included the *Bravo*, the *Guaymas*, the *Acapulco*, the *Agua Prieta*, and the *Tampico*. Like other departments within the military, the navy began building its own equipment in addition to acquiring imports and training its own team of radio technicians and operators.[7]

Radio development in the young air force was surprisingly slow, at least in the numbers of airplanes equipped. During the Calles years, the Department of Aviation operated its own 2,250-watt broadcasting station and had built and distributed a number of receivers. During the early 1930s, the department increased radio and signal education as a regular part of aviator training.[8] It told press agents that equipping planes was a priority.[9] Yet documents from 1929 to 1933 state that only two planes had working radios.[10] That was only one more than existed in 1921, and the same number that existed during the De la Huerta Rebellion. Although these pilots were important to military operations, they apparently had to divide their time between the different conflicts, diluting their impact on the course of the Cristero Revolt and the military rebellions of the late 1920s.

The emphasis on modern war machinery was a global trend, but it was also fueled by events within Mexico. Rebellions continued to plague the country during the last half of the 1920s. The army's leadership believed that radio communications provided a valuable asset in combating rebel groups, which were mostly based in the mountains of western and northern Mexico.

The insurrections that worried political and military officials the most were the Yaqui Uprising and the Cristero Revolt. Many leading state agents considered them interrelated reactionary movements to government policies by misguided "fanatics."[11] Although the Cristero Revolt has received much more attention from scholars, it was the Yaqui Uprising that initially worried Amaro more. The result of a train holdup and years of contention over land invasions and the issue of autonomy, it proved to be last serious Yaqui revolt to date. The renewed attacks against the Yaqui began when a contingent of Yaqui men led by Luis Matus detained a train aboard which Obregón had been riding on September 12, 1926. Calles and the Chamber of Deputies enthusiastically embraced the idea of teaching the indigenous group a final lesson, providing a million pesos to "forever" put down the Yaqui.[12]

Amaro hurled a mass of troops at the Yaqui hideouts in the mountains of Sonora. He sent eight infantry battalions and ten cavalry regiments. Along with President Calles, Amaro worked with General Francisco R. Manzo, chief of operations in Sonora, and General Román Yocupicio Valenzuela, a Mayo Indian who rose in the ranks of the federal military. Yocupicio would join the failed Escobarista Rebellion two years later, surviving to later rule Sonora during the late 1930s.[13] Estimates of the federal forces range from 15,000 to 20,000 soldiers, whereas as Matus led much of his 2,500 followers into the Bacatete Mountains in the Sierra Occidental. Only 1,200 of them were fighters.[14]

The campaign incorporated radio on an unprecedented scale. The army set up three fixed stations in Sonora at Estación Ortiz, Cajeme,

FIG. 19. Soldier working a portable radio, 1931. *Memoria por el SGM*, 1931–32.

and Guaymas. In conjunction with this equipment, six portable radio sets accompanied soldiers in the field. They were used in conjunction with at least one of the planes with a radio device, providing information for reconnaissance, bombing, strafing, and propaganda assignments. Amaro noted that the air force clocked 285 hours on these missions, 100 of them "bombing and machine-gunning."[15] Surrounded by government soldiers, Yaqui combatants and their family members were driven out of their mountain hideouts by the aerial campaigns. Subsequently, the Mexican military dealt the demoralized rebels a critical blow at the battle of Cerro del Gallo in late 1927, more or less ending the rebellion and armed Yaqui resistance.[16]

Similarly, federal soldiers incorporated radio in their response to the Cristero Rebellion. A fixed station and three backpack radios were operating in Michoacán. Another fixed station operated in Zacatecas, which was connected to four other portable devices. Amaro noted in 1928 that there were eleven 7.5-watt portable stations used in operations against the Cristeros, not only in Michoacán and Zacatecas, but

FIG. 20. Military radio operators, station XBH, Mexicali, 1930. *Memoria por el SGM*, 1929–30.

also in Durango, Jalisco, Guanajuato, and San Luis Potosí. That year, the military established another fixed station in San Luis Potosí and built another in Tepic. The station at the Guadalajara airbase and a number of SCOP stations in the Northwest also collaborated with the armed forces.[17]

During the middle of the Cristero Rebellion, a revolt by Generals Francisco Serrano, Arnulfo Gómez, and Héctor Almada over the imposition of a second Obregón presidency showed the world that the Mexican government still faced problems of presidential succession and a military of questionable loyalties. The insurrection, albeit brief, also forced Amaro to pull thousands of soldiers and nine aircraft from the Cristero campaign to search for Gómez, who had fled to the mountains of Veracruz in hopes of rallying a more significant force. This act, in turn, prolonged the Cristero conflict. However, Serrano and Gómez failed to gain a substantial following from within the military, exhibiting, perhaps, that Amaro's reeducation program had actually strengthened allegiance to the state among soldiers. Learning of the rebellion in advance from loose-lipped participants, Calles, Amaro, and Obregón moved quickly to crush the revolt. Immediately, Amaro and Calles "installed their headquarters at Chapultepec castle, where direct wire and radio communications kept them constantly informed." They stayed up late into the night during the first three days of October "snapping" orders.[18] Soldiers captured and executed Serrano in Morelia on October 3. Gómez survived until early November. By then, federal troops under the command of General José Gonzalo Escobar had already defeated his forces. They found Gómez, along with his nephew, hiding in a cave near the city of Xalapa. They too were executed. Almada, who had earlier joined Gómez's forces, escaped to the United States.[19] Despite defeating the rebellion, Obregón was assasinated by a radical Catholic named José León de Toral on July 17, 1928. In his place supporters of Calles and former allies of Obregón agreed that former governor of the state of Tamaulipas Emilio Portes Gil would hold the presidency until new elections in 1930. Calles retained a strong influence on the executive office.

Despite the horrible failure of the Serrano-Gómez Revolt, Escobar and a number of other generals revolted on March 3, 1929. Once again, the issue was the presidential succession. They protested what they saw as the imposition of Pascual Ortiz Rubio's presidency (1930–32). This time the process had been further complicated by the assassination of Obregón, which a number of generals incorrectly blamed on Calles. This insurrection also proved more dangerous, because approximately 28 percent of the armed forces defected to the side of the rebels.[20]

Of course radio, being a regular component of the military and insurrections, continued as an important communications tool during the Escobarista Rebellion. When the uprising first broke out, Portes Gil and Calles learned that General Jesús M. Aguirre, who originally feigned allegiance, was actually in rebellion after José María Dorantos, head of the Seventh Calvary Regiment, sent a radio message to Mexico City stating that Aguirre was really organizing the revolt in Veracruz.[21]

In speeches and publications Portes Gil praised this wireless message as crucial to the government's response.

Although radios were useful, they also had their limitations. Disruptions and interference were still common. Devices broke. A number of them were secondhand. According to a military attaché to the U.S. Embassy who was in contact with a "reliable informant, unusually well situated to acquire the data requested," Mexican agents who purchased radio equipment in the United States in late 1926 were corrupt and bought on the cheap. One such buyer named "Bravo" acquired enough equipment from a military surplus store in New Jersey to piece together twenty-seven transmitters. He bought used equipment so he could pocket money that was provided to him based on the costs of new devices. Officer arrogance also offset the technological advantages of the federal army on more than one occasion.[22] Another limitation was the lack of wireless equipment in airplanes.

In addition to field equipment, the military also increased its use of broadcasting. With the approval of Portes Gil, Julio Tren's Trens News Agency provided bulletins about "the march of military operations against the subversives" on commercial, military, and other government stations. All programs were "favorable to the government and our institutions" and were heard over home and military receivers.[23] Few rebel broadcasting stations existed. Those that did were small, clandestine, and appear not to have lasted long.[24] Insurrections in the late 1920s were at a distinct disadvantage in terms of radio technology.

The government and commercial stations aimed programs at combatants and the general public. The Azcárraga's station, CYL, broadcast anti-Delahuertista news. They disdained Calles's war against the Cristeros, but did not publically condemn it, at least not on the air. Raúl Azcárraga later argued that the government harassed him for his support of Catholic leaders, which was the beginning of the end of CYL.[25] The military provided musical programming and lectures, which were played over SGM stations, the SEP transmitter, and commercial stations. In December 1925, J. M. Puig Casauranc, secretary of public education, told Amaro that he could use the SEP's station to air propaganda and military orders whenever he needed.[26] In 1928, the SGM reported that it operated two 2,000-watt stations "on commercial waves" for communication with boats and planes and for transmitting programs to soldiers and the public at large.[27] Assisted by the military's most renowned radio specialist, José Fernández Ramírez, the Trens News Agency continued reporting on state and military matters into the 1930s, though it was not popular among all military leaders. One officer, Ramón Cortes Gonzales, complained to Amaro in late 1929 that the messages sent out from Balbuena airfield reached Havana and Guatemala, jeopardizing military operations.[28]

Amaro, however, was particularly interested in using broadcasting to shape culture within the armed forces. One of the first subjects addressed by Amaro when he gave his report to Congress in 1925 was his use of radiocasts and receivers as part of a "culturization" program. Similar to a SEP project that was under way, the military program used radio, along with cinema and magazines, to bring expert lecturers to soldiers, discussing leadership, loyalty, Mexican history, and hygiene.[29] Army leaders also hoped that broadcasting would entertain soldiers and raise morale. Station JH operated during 1923 and 1924. During 1925 and the first half of 1926, the Department of Communications and Signals played music and aired lectures from military conferences each Monday via a 6,000-watt transmitter. The station operators temporarily ended their broadcasts on June 1, 1926, in order to help expand radio operations in other parts of the country.[30] In 1929 and 1930, military technicians constructed four short-wave transmitters for military news, educational classes, and general orders.[31] In 1933, the Interdepartmental Radio Commission, a committee comprising specialists who were reviewing the state of radio development in the country, noted that there were two SGM broadcasting stations, XFG and XFH, which aired programs for the armed forces and the public.[32]

In addition to promoting radio within the military, Amaro befriended a number of radio equipment providers. He not only bought radios for himself; he also introduced salesmen to other prominent government officials. In early 1928, Amaro bought two radios, one for himself and one for his wife, from Emilio Azcárraga, the brother of the co-owners of commercial station CYL, and the owner of the Mexican Music Company, a subsidiary of RCA. These devices were not run-of-the-mill radios; they were top of the line. Amaro's machine cost 1,000 pesos. His wife's cost 3,500 pesos (1,000 pesos less than what Azcárraga said he normally charged). Amaro was not the only government official to buy from Azcárraga, and his relationship with state and party leaders, despite possessing misgivings about the government's hard stance against that Catholic Church, grew substantially when he opened his own powerful commercial station—XEW—in 1930.[33] Amaro also introduced "his friend" Hernegildo Robles, a representative of Zenith Radio Corporation of Chicago, to Portes Gil. Robles pressured the president to help the army obtain more apparatuses.[34]

A testament to how important radio was to top military officials, these leaders sent agents abroad to study the medium and to examine how it was used by other governments and armed forces. The army was sending an unprecedented number of people to foreign countries to gain insights into a myriad of subjects pertaining to military modernization.[35] Most of these student representatives went to Europe, but others went to the United States and to South America. For example, in 1928 one mid-level officer—almost all of these officials abroad were young majors or lieutenants—went to Brazil specifically to report on how that country used radio communications. He sent reports back to the SGM, providing information on the developmental history of the medium, personnel, costs, the number of stations, and the number of messages sent and received.[36] Similarly, in 1930 and 1931, another officer sent to Argentina wrote about transatlantic wireless services between Europe and Argentina, radio in Buenos Aires, Argentine-made devices, and communications in Argentina's armed forces.[37] The SGM sent other officials to the Second Meeting of the International Technical Consultative Committee on Radioelectronic Communications held in Copenhagen.[38] In August 1934, the military sent officers training in radio to England and Spain. Back home, the SGM formed a new transmissions battalion.[39]

Military radio was just one example of the serious program of military reform and modernization that Amaro and his successors implemented. Although built on the work done by preceding administrations, they instilled a greater sense of loyalty within the armed forces. Radio played an important role in building this new military culture, putting down the rebellions of the 1920s, and spreading state propaganda to the public, troops, rebels, and foreign audiences.

Public Communications: The SCOP and Private Enterprise

Armed forces dominated wireless use during the Revolution, but as addressed in chapter 4, radio communications expanded into a number of commercial enterprises and other state agencies during the Obregón and Calles years. Indeed, radio use among commercial aviators and merchant sea vessels outpaced the military by the early 1930s. Radio messaging from trains, first started by factional leaders during the Revolution, became more commonplace on regular passenger services. As a result, radio became an integral part of communication in transportation, advancing the practice of using the medium to interlink the various parts of the country. The SCOP, which had been a major player in Mexican radio development from the beginning, oversaw these developments and provided permits to businesses interested in using wireless technology. From 1926 to 1935, SCOP agents reported that the percentage of all wireless messages received and transmitted rose from 8 percent to over 14 percent.[40] Radio had become a vital part of government and public communications.

The DGTN within the SCOP continued to direct much of the international radio services. Stations constructed in Central America during the Carranza and Obregón years kept the government leaders of El Salvador, Guatemala, and Costa Rica in direct contact with Mexico City. In 1926 the Mexican government worked out a number of new wire and wireless contracts with Central American and Caribbean countries.[41] The DGTN also expanded its communications with nations in Europe, Asia, and South America. The Chapultepec radio towers continued to beam government newscasts throughout the latter half of the 1920s and the 1930s to domestic and foreign stations. In fact, the station increased its power dramatically. In 1929, the SCOP completed a massive receiver in its newly founded Palo Alto station in the Lomas de Santa Fe region of the Federal District. The operation picked up radio messages from around the entire globe.[42] In addition to using radiotelegraphy, DGTN employees incorporated international radiotelephone services starting in the early 1930s. In January 1931, SCOP director Juan Andreu Almazán, later to play a leading role in galvanizing much of the Mexican right against Cárdenas, talked via a wireless "phone" with Primo Villa Michel, the Mexican ambassador in Germany.[43]

During the presidency of Pascual Ortiz Rubio, the SCOP allowed a number of U.S. providers to start wireless operations in Mexico. These agreements, however, had a large number of stipulations to alleviate fears of foreign domination. For example, in 1931 the SCOP signed a deal with the Montgomery Company, a subsidiary of Frutera Transcontinental, which was, in turn, controlled by United Fruit. The concession allowed for the Montgomery Company to start a wireless office in Ixtapa, Jalisco, close to the business's banana plantations in the vicinity of Balderas Bay, 200 miles south of Mazatlán. United Fruit had long been one of the main radio innovators and developers in Latin America, but it had to concede to a number of strict stipulations in order to operate in Mexico. For one, the station had to be accessible to the public and interlinked with the national communications network, though preference was to be given to the company's messages. The station also had to hire and pay for a Mexican staff, including an operator, a mechanic, and a messenger. The government retained the right to close the operation at any time and to buy the station if it chose to do so after ten years. Additionally, the company agreed "not to invoke the aid of their diplomatic representative" or the contract would become null and void and the installation would become "property of the nation."[44] Agreeing to the terms, the Montgomery Company used the station mostly to communicate with sea vessels involved in transporting their products or with other ships and stations connected to United Fruit. It appears that the operation also kept in touch with certain SCOP stations. Indeed, the Mexican government specified which stations it could contact, including the offices in Chapultepec and Mérida within Mexico; Barrios, Guatemala; Tela and Tegucigalpa, Honduras; Managua, Nicaragua; and New Orleans and Miami in the United States.[45]

On January 14, 1932, the SCOP and RCA signed a ten-year contract creating Radiomex, R.C.A., which provided more wireless connections to New York, Berlin, Madrid, Havana, and starting in 1934, Tokyo. Both RCA and the government shared the profits and costs. SCOP officials operated the Mexican end of the communication link via Chapultepec, and RCA employees operated traffic to and through the United States via their operation on Long Island. The agreement, decreed into law by Ortiz Rubio on June 17, 1932, illustrates the decision by top government officials to allow foreign companies a larger role in international communications to and from Mexico, conceding the United States' dominant role in global wireless technology and know-how.[46] This trend continued during the Cárdenas administration.

State and local governments also operated their own radio stations. In 1924 Chihuahua's Department of Telegraphs started XICE, a small operation that broadcast every Wednesday and Friday.[47] The next year D. Javier M. Eroza, the mayor of Mérida, Yucatán, installed a radio station on property controlled by the Central Resistance League.[48] While governor of Veracruz, General Heriberto Jara inaugurated the state's own station in Xalapa in late 1926.[49] According to a 1933 report, the state government of San Luis Potosí operated four radio stations, Tabasco had two, and Chiapas relied on one.[50]

Radio also took to the skies—at least outside of the military—in much greater numbers during the late 1920s and 1930s. Passenger and mail aviation services relied heavily on the new technology. Of course, the SCOP had to approve all of these operations. Many of the pilots were former or active military aviators. In January 1927, pilots of the National Air Force scouted routes that could be used for government and commercial purposes.[51] In April, the SCOP approved the first postal air route from Mexico City to Túxpan to Tampico. Another route began that year between Mexico City, Veracruz, Puerto México (today, Coatzacoalcos), Salina Cruz, and, finally, Tapachula, Chiapas.[52] The same year, the SCOP also approved eighteen other permissions for aerial navigation, including three to the Mexican Aviation Company. One of its owners was none other than Enrique Schöndube, the German immigrant and agent of AEG and Telefunken that engineered Mexico's first public wireless stations.[53] Over the next few years, a number of companies began domestic and international services, especially to the United States. The most successful air services relied on radio for communications. By 1933 the Mexican Aviation Company possessed twelve radio stations: Tampico, Tamaulipas; Rancho Méndez, Hidalgo; Mexico City DF; Túxpan, Veracruz; Tejeria, Veracruz; Minatitlán, Veracruz; Villahermosa, Tabasco; Ciudad de Carmen, Quintana Roo; Campeche, Campeche; Mérida, Yucatán; Payo Obispo, Quintana Roo; and Tapachula, Chiapas. The Central Air-Ways Company had stations in Mexico City DF; León, Guanajuato; and Torreón, Coahuila.[54]

In 1929, President Portes Gil talked via radiotelephone on his "presidential plane" during Mexico's Air Week celebrations. In addition to speaking to the president of the Mexican Aviation Company, the president sent a message to his wife: "From an altitude of 23,000 feet, as we are now over the crater of El Popo the volcano and enjoying the gorgeous beauties of nature, to you, to mother, and to Chacha, our little girl, loving greetings.—Emilio."[55] Radio had become a regular part of travel by plane, further expanding the diversification of electronic wireless communication and its use in transportation.

Radiotelephone services, including those owned by the Mexican Aviation Company, were not confined to communicating with planes. Radio operators from this company spread news of a Salvadoran coup that it overheard on its receivers. Some of the first reports of murders, the disposal of President Arturo Araujo, and the otherwise "calm situation" heard in Mexico and the United States came from the company.[56] Radio messages entered and influenced nations outside of their place of origin, and the importance of the medium to the foreign relations of nations should not be underemphasized.

Other transportation sectors incorporated radio for public use. Starting in 1928, the DGTN established equipment and personnel to provide wireless services aboard a number of passenger trains that serviced the states of Coahuila, Chihuahua, San Luis Potosí, Tamaulipas, Nuevo León, Querétaro, Guanajuato, Aguascalientes, Sonora, Sinaloa, and Jalisco. According to one Mexico City newspaper, passengers needed only to grab the attention of a DGTN agent while on the locomotive and make a request.[57]

In addition to public railway services, SCOP officials made sure that the presidential train had working radio equipment. The Yellow Train used by Obregón and Calles possessed wireless equipment, much like that Villa used during the Revolution. In 1925, the Calles administration invited journalists to ride on the presidential train to see where "grand irrigation works" were soon to be built in Chihuahua. The SCOP operators helped them send reports via radio back to their respective newspapers in Mexico City.[58] Two years later, a number of government officials decided that a new train was in order. Through Arturo M. Elías, the president's half-brother and the Mexican consul-general in New York, the government purchased a truly opulent train from Chicago's renowned Pullman Company. The Olive Train consisted of five luxurious coaches for Calles and his staff and another coach specifically for the SGM. The machine cost well over US$500,000 and possessed "every conceivable necessity," including bulletproof cladding and "the most modern radio sets," which were in every coach excepting the baggage and escort cars. The *Los Angeles Times* called it the "most palatial train in the world."[59] A year later, the SCOP added a new short-wave radio built in the Chapultepec workshop.[60]

Even before the 1930s, radio had become an important component of travel, specifically for those that could afford the luxury of trains and airplanes. Wireless communications allowed for better coordinated military and political actions, and additionally became a part of journalism, business, and personal correspondence. Like the irrigation projects in Chihuahua, radios themselves represented Mexican modernity, connecting the Revolution to promoted ideals of progress.

Foreign Relations and Legislation

As radio became more prominent, executive officials and congressmen signed new legislation regulating the medium. Diplomats also made new agreements on communications with a number of other countries, while arguing with the United States over the role of the state in broadcasting, the allocation of radio frequencies, and interference caused by each other's transmissions. Debates over these issues, which increased during the last year of the Obregón presidency, continued with greater frequency during the last half of the 1920s and the 1930s, straining U.S.-Mexican relations and influencing the contours of Mexican domestic policies.

As stated in chapter 4, the Obregón administration took up a prominent role in the international debate about radio development when it hosted the Inter-American Committee on Electronic Communications in Mexico City in the summer of 1924. The event became a showdown between the United States and Latin America over the role of states in radio development in the Western Hemisphere. Mexican officials, who had striven to increase their own influence in Latin America, especially Central America, had attempted with some success to convince many of the region's leaders that they should build nationalist wireless systems, operated by their respective governments, and not to rely on U.S. companies like RCA to provide the service. South American countries, including Brazil and Argentina, pushed similar views, as they had already shown in the Fifth International Conference of American States held in Santiago, Chile, in 1923.

A number of conferences, both global and regional—Mexico, United States, Canada, and Cuba—followed over the next decade. The 1927 International Radiotelegraph Convention of Washington, and the 1932 International Telegraphic and Radio Telegraphic Conference in Madrid were significant, but other bilateral agreements and smaller regional meetings proved just as important. Most Latin American countries, including Mexico, continued to operate a number of their radiotelegraphic services through the government; however, many Latin American representatives became more willing to cooperate with U.S. private businesses, which dominated radio manufacturing and technological innovation. Delegates at these conferences worked out the details for radiotelegraphic and radiotelephonic messages for maritime and airplane communications, meteorology, scientific experiments, warfare, and commercial and state broadcasting. The 1927 conference delegates also agreed to change the call letters of the participating countries, providing Mexico's final XAA-XFZ designation.[61]

One significant impact of these international agreements is that they prompted Mexican officials to increase and to modify their own legislation. For example, the 1924 Mexico City meeting set the foundations for the 1926 Law of Electronic Communications.[62] The main components of the law came directly from the positions offered by the Mexican delegation at the Mexico City conference. These included strict government supervision over private wireless operators, and the requirement of citizenship for radio-station owners. Both the conference and subsequent legislation also demonstrated that national security risks stemming from the Revolution and the De la Huerta Rebellion remained a prominent concern of communications officials. Delegates ardently pushed for the state's "right to seize such stations or suspend their service in case of local or international disturbance" and for "all parties concerned [to] keep secrecy in respect to all messages or information intercepted by receiving apparatus not intended for the general public or for the party or parties hearing the same."[63] The 1926 law also made it illegal to report news contrary to the security of the state, international harmony, public order, good customs, decent language, or "attacks of any form against government or individuals."[64] However, showing the growing acceptance and outright promotion of commercial broadcasting stations, government officials increased the possible length of a station concession to fifty years.[65] In other words, Mexico pursued a mixed commercial-government system that promoted private broadcasting but with strict state oversight.

The focus of most of the discussion between the Mexican government and U.S., Canadian, and Cuban officials revolved around wave-frequency allocations and the related issue of interference. Radio transmissions broadcast from different locations that share the same or similar wavelength interfered with each other, often making signals unintelligible; a well-known fact that was used to jam signals during World War I and the Revolution. The expanding number of commercial and amateur broadcasters in the hemisphere crowded the airwaves, causing constant interference problems. Certain wavelengths also carry sound clearer and longer than others. Broadcasting businessmen who operated powerful transmitters coveted these frequencies, known as clear channels.

During the 1927 Washington DC, conference, American delegates introduced legislation on wavelength allocation and interference, which ultimately became Articles 5 and 10 of the resulting document. These measures stated that people should open new stations only if they do not interfere with others. Since the United States operated the vast majority of stations, including those on clear channels, they, by means of earlier and quicker development, dominated the airwaves, legally leaving few to others.[66]

Despite agreeing to the articles prohibiting new stations from using frequencies that would interfere with already established American broadcasters, the Mexican government, starting in 1930, allowed the construction of massively powerful radio stations along the U.S. border that operated on frequencies extremely close to commercial operations in the United States.[67] The government assisted with their construction.[68] Medical charlatans and other ostracized Americans chased off

FIG. 21. John Brinkley's radio station XER, Villa Acuña, c. 1931. Courtesy of Library of Congress, Prints and Photographs Division, Reproduction No. LC-USZ62-97961.

the airwaves in the United States operated these stations, transmitting a motley assortment of Latin American and U.S. music, fortune tellers, and advertisements for outrageous medical cures. The most notorious "border blaster" of the 1930s were John R. Brinkley—known as the "goat-gland doctor" for his procedure of rejuvenating male sexual performance and fertility through sewing pieces of goat testicles into men's testicles—and Norman Baker, who peddled false cures for cancer.[69]

Understandably, these operations infuriated a large number of U.S. radio-station owners and listeners, whose programming suffered from terrible interference. The border blasters also irked U.S. medical professionals who had worked hard to delegitimize Brinkley and Baker. Other listeners, however, enjoyed the eccentric and eclectic entertainment they provided. Brinkley's station received hundreds of thousands of positive letters from all over the United States, especially from the

Midwest. Indeed, the rise of American folk and hillbilly music in the United States owes much to these powerful frontier stations that aired transmissions of Woody Guthrie, the Carter Family, and other musicians. Sadly, thousands of seriously ill people also believed in their false cures.[70]

A number of prominent Mexicans, in addition to Americans, opposed these stations. Just as in the United States, many medical professionals, especially those involved in the Department of Public Health, despised Brinkley and Baker's quackery. Lawyer Manuel Ruíz Sandoval protested to Coahuila's Governor Nazario S. Ortiz Garza that Brinkley and the other American broadcasters operating in Mexico were agents of U.S. imperialism. Americans found the claims ridiculous. One opinion piece in the *Los Angeles Times* facetiously quipped that "according to all photographs, the doctor's hirsute embellishment is a Vandyke not an imperialist."[71] Apparently to this opinion writer, Brinkley's facial hair showed that he was more selfish and pompous than imperial. But to some Mexicans who took the jingoist campaigns of the newly formed PNR seriously, these stations, approved by members of the very same party, flew in the face of their own touted nationalist policies.[72]

Nonetheless, to many PNR and government members—and the distinction between the two was increasingly slim—these stations proved useful. As a result, state officials often ignored their illegal trespasses, though they occasionally hit the stations with payable fines and for bribes for themselves. These gringo station "owners" (on paper they designated in-name, local partners as owners to get around laws requiring Mexican ownership) also lined the pockets of local authorities and employed Mexicans in a wide number of capacities.

The stations, however, provided much more than the occasional kickback. Although not exactly intuitive, the stations, by blanketing the United States and parts of Canada, became the biggest and most effective advertisers of Mexican tourism and culture, which had become in vogue among administrators as a way to increase the struggling economy already in the pangs of the Depression. Indeed, some border stations offered free airtime to the Mexican government to promote their "nationalist" tourism campaigns. There was no doubt that these American radio rebels were stimulating the border economy. Radio performers and their families spent money in local stores. Brinkley bought the whole Villa Acuña police force new uniforms. Station managers hired locals for construction jobs. Radio charlatans, local businesses, and government officials alike cashed in on radio and America's growing obsession with auto travel and "all things Mexican."[73] These stations also became the government's greatest weapon in a battle to obtain more of the clear channels that U.S. stations hoarded and jealously guarded.[74]

Legislators did try to counter criticisms levied against the border blasters by introducing new laws, even if officials only enforced them when they wanted to. Some of the most considerable components of these regulations show the government's increased demands of commercial-sector cooperation in promoting revolutionary credos and state-sponsored culture. Under new rules created in the 1931 General Means of Communication and Methods of Transport Laws and the 1932 General Means of Communication Law, commercial stations had to provide airtime to the Department of Public Health and other state agencies. Broadcasting religious messages—and fortune telling—became illegal, and the same national security caveats were reinforced. Nationalist elements also figured prominently. Station owners had to possess a staff that was 80 percent Mexican. Another article, aimed at border stations, outlawed the practice of having operations in foreign countries telephone or radio in shows to be rebroadcast from Mexican stations without government permission. Additionally, broadcasters had to speak in Spanish unless they obtained permission to do otherwise. The 1932 law went so far as to make it illegal to use consistently bad grammar that insulted the "purity" of the Spanish language. Any violation of these or other infractions resulted in fines added to the already-existing 5 percent tax that the government took from all station profits. In 1933, however, new regulations made it legal to broadcast in another language without government permission as long as the programs were provided in Spanish first and then translated. Something border blasters still regularly failed to do.[75]

In international circles, the issue of awarding wave frequencies based on primacy of use arose again in the 1932 International Telegraphic and Radio Telegraphic Conference in Madrid. Hoping to avoid the issue while remaining an active participant in the conference, the Mexican delegates approached the Americans on the first day of the conference and put forward a plan to work out their differences on clear channels during a smaller regional meeting the following year. Eager to retain their dominant position in communications in the Western Hemisphere and to limit the influence of Europe, the Americans agreed. In fact, U.S. officials had worked out a bilateral agreement over clear channels with Canada just before the conference in Madrid.[76]

The 1933 North and Central American Regional Radio Conference took place in Mexico City. As its name suggests, it included other nations in addition to the United States and Mexico: Canada, Costa Rica, Cuba, El Salvador, Guatemala, Honduras, and Nicaragua also sent delegates. The issues of clear channel distribution, border blasters, and interference in general dominated the convention. Immediately preceding the gathering, observers in Mexico and the United States believed that some agreement over frequency redistribution and the border stations would be reached. As a sign of cooperation, Mexican officials enacted the 1933 regulations, aimed largely at the border blasters, just before the meeting. The SCOP also hit Brinkley's operation with a new round of fines that reached upward of US$150,000.[77] However, the American delegation responded in Mexico City with arrogance. According to Josephus Daniels, a former U.S. secretary of navy involved with previous radio disputes, U.S. representatives offered an "impossible ultimatum which the Mexicans understood as an attempt to dictate what they should do with stations in their country. The Americans virtually demanded that Dr. Brinkley be put off the air and insisted that Mexico should have no stations on the border that could carry messages into the United States. This demand made failure certain from the beginning."[78] In retaliation, Mexican delegates discussed upping the power of Brinkley's station with former U.S. vice president Charles Curtis, who worked in the interest of XER. They also considered granting permits to establish two more frontier stations.[79]

Most American broadcasting company executives sighed with relief as the discussions broke down. They feared any redistribution of the coveted frequencies, of which Americans possessed ninety of the ninety-six available wavelengths. Mexican representatives had wanted six of the channels, the same number that the Americans had provided to Canada in their bilateral agreement the year before. The U.S. delegates offered three and the conversation stalled. The Americans also refused to give any of the contested wavelengths to the other participants, who also clamored for what they considered their fair share of the radio spectrum. Similarly, the delegates failed to come to any agreement over the border stations. The conference did, however, make progress on eliminating interference on point-to-point radio communications, which operated in a different frequency range.[80]

By 1934, U.S. radiomen in Mexico had erected at least five powerful stations along the border. On top of those five, Americans citizens were in the process of having more built. That year, however, President Abelardo Rodríguez's administration (1932–34) moved in the other direction, shutting down Brinkley's station XER for violating the 1932 and 1933 rules on language and religion. Relatedly, the PNR took over the station, conveniently making it a part of the party's recently established broadcasting chain. Once again state leaders used these border stations, built with American capital, to their advantage. The other stations continued, despite breaking the same rules. U.S. legislators tried to impede these operations by enacting the 1934 Communications Act, creating the Federal Communications Commission (FCC) and banning the use of transnational remote-control radio operations. The FCC became a permanent fixture in the United States, but the law did little to slow the frontier operations. High-powered "outlaw" stations operated on the Mexican side of the border into the 1980s, when the

Mexican and U.S. governments finally came to an agreement on clear channels.[81]

Conclusion: Radio Control

Despite the massive changes in radio technology during the 1920s and 1930s, control remained the prominent issue among state leaders. While commercial broadcasting became a firmly established part of Mexican society, the military continued to use radio to combat a number of rebellions, culminating in the Escobarista Rebellion of 1929. Unlike the previous De la Huerta Rebellion, these revolts mostly failed to seriously jeopardize the state's grasp over the national wireless network. In conjunction with an ever-increasing number of field radios, the armed forces under Calles, Amaro, and the PNR monopolized military communications and used them extensively against their enemies. Radio was an important tool that, in addition to new weaponry and airplanes, allowed the post-1930 armed forces to maintain power. Just as importantly, broadcasting helped increase the level of military allegiance to the president and the governing party within its ranks. Radio proved useful on the battlefield, in the barracks, and in the home.

In addition to radio's importance to the military's ability to suppress, or at least contain, revolts, the technology proved crucial to the country's structural development. Wireless devices significantly improved the service of passenger airplanes and railways in the late 1920s and early 1930s. Radiotelegraphs and radiotelephones also brought some alleviation to the overburdened wire system. By the mid-1930s, radio had become a regular component in international communications for the state, commercial enterprises, and private citizens. To keep up with the fast pace of technological change and to improve the utility of radio operations, the SCOP increasingly turned to foreign companies, especially RCA, in order to better reach international audiences. Mexican officials, however, still dominated domestic services. At the height of Mexican nationalism in the 1930s, radio had surprisingly become less directly state controlled than ever despite the government's increased involvement in broadcasting. Nevertheless, state leaders kept in place a large number of nationalist restrictions, and the SCOP continued to control the vast majority of the internal communications network.

The partial shift to commercial radio providers correlates with the decision first made by the Obregón administration to pursue a mixed state-private broadcasting system. While state officials in a number of departments started their own operations, the government focused just as much attention on protecting domestic commercial radiocasters. They did this with considerable success, restricting the dominance of American radio corporations, which even when affiliated with Mexican companies, controlled little in regards to content and day-to-day operations. The border blasters provide the one glaring exception. Government leaders tolerated these renegade American-Mexican operations as long as they proved useful as a bargaining chip in international communications forums, for promoting tourism, and lining the pockets of officials. But as the Brinkley station exemplified, government toleration of abuses was limited, especially when the state party decided it wanted a station for its own purposes. Relying on foreign businesses for radio equipment while restricting outside control of communications services remained a prominent aspect of Mexico's wireless policies, which shows a great deal of consistency with previous decisions made during the Porfiriato and the Revolution.

One change was the inclusion of greater commercial involvement. Government leaders reinforced their relationships with wealthy commercial broadcasters as a means of amplifying nationalism and allegiance to the one-party political system established in the early 1930s. This process, in turn, dovetailed with monopolistic tendencies in political and commercial broadcasting, curtailing and co-opting the more democratic tendencies of amateur radiocasters, who possessed much weaker equipment.[82] Chapter 6 addresses this subject, focusing predominately on the same period, but giving more consideration to cultural broadcasting, populist politics, and the foundations of the single-party state.

6 Broadcasting State Culture and Populist Politics

> The history of radio cannot be told simply in terms of devices, inventors, and manufactures, but must be integrated with the history of political power and information.
>
> Daniel Headrick, *The Invisible Weapon*

On November 30, 1924, even more people filled the streets of Mexico City than usual. Sounds of cars and carriages shared the air with the aromas of food vendors and human settlement. Fifty thousand eager spectators packed themselves into the National Stadium to witness a historic event, the first peaceful transfer of the presidency since 1884. In honor of the incoming and outgoing presidents, a folkloric ballet consisting of five hundred different couples, danced *jarabes tapatíos,* or hat dances.[1] Calles, with Obregón at his side, addressed the massive crowd, and the entire country. This occasion provided the opportunity for the first broadcast of a presidential inauguration in Mexican history.

As radio became more popular in the late 1920s and 1930s, Calles and his allies increasingly depended on it not only as a military tool but also to spread nationalist propaganda that equated state institutions—and starting in 1929 the PNR—with the culmination of the ideals fought for during the Revolution. Broadcasting held great potential for building a wide base of support. It also fostered a partnership between members of the often-adversarial commercial and government elites. In exchange for carrying political messages and complying with state regulations, government leaders rewarded the owners of the largest commercial stations with favoritism. Radio helped facilitate a loose, cross-class alliance.

Broadcasting had become an essential component of the controlled incorporation of the general population into politics. Government leaders used the medium to turn campesinos, laborers, businessmen, and middle-class urbanites into state allies, with some success. Although many areas remained out of the constant reach of the government, or stood opposed to its intrusion, the voice of state representatives reached greater audiences than ever before. For good reason, scholars have painted Lázaro Cárdenas, who took office in 1934, as the poster child of government broadcasting and populist politics, but by the time he took office presidents had been using broadcasting for over a decade and wireless telegraphy for thirty-five years. The traditions of radiocasting inaugural addresses, talks to and by Congress, New Year's Day speeches, presidential campaigns, and during national crises, had all been a part of the broadcasting system built by associates of Obregón and Calles, the PNR, and private-station owners. Cárdenas may have been more radical, but he built squarely on well-established foundations.

In the process of constructing populist coalitions, Calles and top leaders of the PNR, including Cárdenas, also used radio to dominate over smaller regional parties. By monopolizing political speech on the airwaves, the PNR provided the furthest reaching propaganda, reinforcing its prominent position. In addition to owning the most powerful party station, PNR members provided radios to labor unions and rural community centers, incorporated political broadcasting into education, and obtained the cooperation of stations run by regional parties, eventually subsuming them. In addition, the SEP, Chapultepec, army, Secretariat of Industry, Commerce, and Labor (SICT), and SCOP all transmitted pro-PNR programming. At its high mark in 1938, fourteen government stations existed.[2] State agencies also broadcast on commercial stations, which by law, had to allow airtime for state bulletins and to play "Mexican" music. The PNR gained the cooperation of the most dominant commercial broadcasters, building a close relationship between media moguls and a one-party system that would evolve for generations.

During the Calles era, state agencies expanded their direct involvement in broadcasting. In a time when legitimacy was associated with the keeping of long-standing promises to "the people," broadcasting provided a way for the government to persuade citizens to join its causes. Through radio stations—especially CZE/XFX of the SEP, CZI/XEFI of the SICT, and XEO/XEFO and XEUZ of the PNR—the Calles administration and the subsequent governments that ruled under Calles's influence established the medium as a pillar of populist politics and, after 1929, of the one-party system first represented as the PNR.

Radio Instrucción Pública

The government's greatest direct effort in broadcasting from 1924 to 1931 was through the SEP, created in 1921 by President Obregón and José Vasconcelos, the renowned scholar and the department's first director. Its station, CZE (changed to XFX in late 1928), reached thousands of Mexicans in addition to listeners abroad. During an address to the National Congress in September 1924, Obregón laid out the goal of this new endeavor: "Teachers will meet somewhere near the ranches and neighborhoods of their students to transmit, on a predetermined day and time, a lesson about a useful theme, music, and news to arouse their interest so they can participate in the life of our country."[3] The following year President Calles created a specific department—Radio Educational Outreach—to tackle the daunting task of getting receivers into rural schools.[4]

The use of radio coincided with the SEP's cultural missionary program. State leaders hoped that secularizing evangelists would educate Mexico's rural, indigenous populations, creating more productive citizens with a stronger allegiance to the revolutionary state. Indeed, the increase in the central government's control of provincial education was one of the hallmarks of the SEP, just as mass politics became a consequence of the Revolution in general. Although originally limited to states close to the capital, such as México, Hidalgo, Puebla, Tlaxcala, and Morelos, Calles expanded the program to other areas by 1926. By the end of the Calles era, the program had reached at least twenty states.[5] Radios were important tools in this modernizing effort. Prominent educational leaders believed radio to be "one of the best modern advances to diffuse education and culture."[6]

The station's early broadcasts mirrored the agendas of the SEP leadership, starting out conservative and elitist, but becoming broader in scope as policies changed and community input influenced programming. Shows originally included classical music and scholars who discussed a myriad of academic topics in a "simple and plain language."[7] The department's first directors—José Vasconcelos, Bernardo J. Gastélum, and José Manuel Puig Casauranc—hoped to bring good European tastes, nationalist composers, and the best Mexican experts into people's homes and schools.[8]

CZE also provided programs designed to specifically "arouse the interest of the Indian . . . native orchestras playing ancient native airs and songs in native dialects." Of course, no record of "ancient native airs" existed, so the directors of these ensembles surely had a lot of creative leeway. Lectures followed the songs, sometimes in indigenous languages, "dealing with the needs, aspirations, and opportunities of the Indian races."[9] Topics included geography, natural resources, hygiene, family finances, and home improvement. Ethnographic discussions about Indians themselves were also fairly commonplace, providing metropolitan perspectives on the lives of the radio recipients and their neighboring communities, whom they had just recently studied. Indeed, the growing influence of anthropologists in the country had a significant impact on the cultural missionaries, who also often acted as amateur ethnographers, providing material for later radio programs. Some of the 1926 broadcasts included "The Indigenous Tribes in Mexico," "Our Archeological Riches," and "Natural Hygiene in Rural Villages."[10]

During the late 1920s, the SEP's station managers strove to expand the listenership to all "radiofiles" in Mexico, competing directly with popular commercial and amateur stations. To persuade villagers to listen to CZE/XFX, the SEP locked tuning dials on donated radios to the station's wave frequency. However, radio inspectors regularly found that villagers broke the seals and listened to domestic and foreign commercial programming.[11] By 1928—the year CZE became XFX—the station diversified its programming. In addition to educational talks, weather reports, and Chopin, the station started providing a variety of Mexican folk songs and even blues and jazz. The Jazz Band of the Traffic Department, a regular fixture in 1928, played songs including "Mi Regular Gal" and "Weary Blues."[12] Guty Cárdenas, one of Mexico's first international radio stars, played his increasingly famous pieces in addition to renditions of songs from his home state of Yucatán. Other music celebrities, such as Alfonso Esparza Oteo and Guillermo Posadas, worked popular Mexican songs into big band arrangements.[13] As commercial stations were vying for government support by producing nationalist programming, the SEP station competed with the private broadcasters by incorporating more popular music. By 1930, however, Alejandro Michel, the Director of Radio Educational Outreach, reported that as part of a "nationalist labor" campaign—a joint endeavor of the SEP and the newly formed PNR—he wanted "to counteract the effects of American jazz" by playing Mexican regional and popular music in a show that aired every Thursday and ended with the national anthem.[14]

Most villagers introduced to radios found them amazing. For many people it was the first time they had heard a voice transmitted through a machine. SEP missionaries explained that Indians and campesinos, young and old, reacted enthusiastically to the "mysterious" and "magical" devices.[15] When educators brought the receivers to villages, the community usually put on a large celebration. In Acatlán, Puebla, one such party included orchestral performances, poem recitations, a tango, inauguration speeches by community leaders, an address by SEP zone inspector C. E. Sansalvador, and lastly, listening to the radio.[16] In other words, it was more often than not the technology itself that awed rural residents, not any specific message or program.

Journalist Fernando Ramírez de Aguilar, who often used the pen name Jacobo Dalevuelta, made a similar statement about his experiences in small pueblos with radio receivers: "It is very curious to observe the small town away from the metropolis where they have a receiver, the enthusiasm with which every night around the apparatus the young and old wait for the remote and mysterious words that magically comes from the loudspeakers."[17] Much the same as Mexico City urbanites who encountered radio for the first time at the 1923 Mexico City Grand Radio Fair, most campesinos introduced to broadcasting found the medium, and the radios themselves, captivating.

Though many rural community members welcomed the SEP and their machines, others resented the intrusion of the federal government. As one author noted, sometimes teachers "found it necessary to wear side arms and to sleep with a rifle next to the bed" because of stiff resistance to state education and government policies against the Catholic Church.[18] A number of teachers were killed, especially during the widespread Cristero Revolt. Even afterwards, violence continued into the 1930s. In 1933, in the small town of Tonantzintla, Puebla, radio inspector Luis F. Rodríguez found that the donated device worked perfectly, that teachers used it to listen to XFX, and that the station was having an appreciable impact by influencing teaching curriculum. However, half of the town's residents refused to listen to the machine because of disagreements over the policies of the government, causing a rift in the community.[19]

Much of the controversy revolved around educating children. During the late 1920s and early 1930s, XFX offered a number of programs specifically for kids. According to historian Elena Jackson Albarrán, programming on XFX assisted the agency's project to influence children by making history and technology more relevant to them and inspiring social action.[20] These programs included regular academic subjects that federal teachers had been trained to teach. For example, the radio series *Periódico Infantil*, which broadcast once in the morning and once in the afternoon starting in 1930, provided brief lessons on the history of the Revolution. Short lectures on zoology, botany, physics, history, geography, or the "national language" followed.[21] XFX's singing lessons proved to be one of the station's more popular programs in schools. In the communities with functioning receivers, inspectors reported that the teachers often used the show in their classes.[22] Another program developed in the early 1930s was *Troka the Powerful*, created by Estridentista writer and artist Germán Liszt Arzubide. It taught children the benefits of machines and modernization through the narration of a somewhat terrifying robot figure. Yet another show, *Antena Campesina*, focused similarly on the benefits of modernization. It, however, was directed more specifically at indigenous children and their mothers.[23]

Outside of students, CZE/XFX targeted mothers and housewives. In 1929, SEP officials bragged that the station reached over three thousand housewives daily, who listened to productions including "How to Make a Practical and Economical Menu" and "How to Become a Good Housewife."[24] In the 1930s, the station regularly carried a program called *The Home Hour*. Much of the early-hour broadcasting focused on cooking, beauty, budgeting, and general "domestic science" issues. As communications scholar Joy Elizabeth Hayes aptly put it, "Educators saw radio as an effective way to penetrate the home: to consolidate the modern, secular housewife and mother and the developmentalist state."[25] As the disseminator of family values in the home, state officials hoped that mothers would influence their children and husbands.

The SEP station tried to appeal to rural fathers as well, but largely through evening broadcasts. They focused more on agriculture, livestock, small industries, geography, travel shows, the weather, music, and history lessons.[26] In 1931, XFX broadcast a program specifically on manual labor for rural men.[27] As Undersecretary of Public Education Moisés Sáenz stated in *Mexican Folkways* in 1928, the SEP wanted to make "the school the home of the village," the new church, the new community center.[28] Along with other educational leaders in Mexico City, he partnered with allied municipal officers to get campesinos into schools to listen to radio broadcasts in hopes of improving their crop yields and their abilities as manual laborers. Sáenz additionally wanted residents to learn basic economics and align themselves with the national government.

But did campesinos actually listen to these programs? When Calles took office, he commented that the government had succeeded in placing radios in the majority of federal schools, which numbered approximately one thousand.[29] Calles and the officials informing him surely exaggerated. Over the next five years, other government representatives continued to talk of placing thousands of radios in schools as well. In Veracruz in late 1929, radical governor Adalberto Tejada and Director of the Department of Universities Genaro Angeles partnered their efforts through Jalapa station XFC with the SEP in order to expand political and educational programming in their state. According to SEP officials, this collaboration allowed the "poor people of the state of Veracruz, part of Tamaulipas, Puebla, Zacatecas, and Aguascalientes" to benefit from educational broadcasting.[30] Tejada and Angeles told federal officials and the Veracruz newspaper *El Informador* that they were going to equip all of the approximately two thousand schools in their state.[31] But even with the help of local pro-radio committees in small towns including Coatepec, Tecelo, and Xico, it appears that they never closed in on their mark.[32] After receiving a substantial increase in federal funding in 1928, Minister of Public Education Ezequiel Padilla also claimed that he would significantly increase the amount of receivers in the growing number of public education institutions.[33]

Despite the claims about thousands of radios, it remains difficult to get an accurate count on the exact number of the devices donated or sold by the SEP to schools. According to SEP records, newspapers, accounts from the U.S. Department of Commerce, and lists by other scholars, it appears the number was probably in the low thousands for the entire ten-year period of 1924 to 1934.[34] The biggest year for radio placement looks to have been 1930, when the U.S. Commerce Department stated that seven hundred receivers had been given out to schools across Mexico.[35]

In addition to foreign-made devices, the SEP prided itself on providing domestically built radios, fashioned by educators, students, and employees of the ministry's Technical Section. A number of the radios used in SEP's rural school campaigns were Mexican-made Titlanti models. Originally, SEP officials donated these apparatuses at no cost to the schools. By the late 1920s, however, they charged forty pesos—more or less the production cost—which members of villages had to pool together in order to acquire the apparatus.[36] It is unclear how many of these devices were made, but at least dozens, possibly more.

Donations were another source of radios. Multiple companies gave the SEP receivers, especially after the ministry's Radio Department began a campaign that traded airtime for commercial advertisements in exchange for the devices. In the last half of 1928 alone, the SEP obtained twenty-five radios by this method.[37] By 1933, however, the Office of Cultural Radiotelephony was turning down requests from rural schools for radios. Budget cuts resulting from the Great Depression had halted the spread of the program. Instead, officials worked on repairing and maximizing the use of the radios already in operation.

Another important factor in obtaining at least a qualitative conclusion on the reach of these programs in rural areas is to determine if people in communities with radios actually listened to them. All evidence shows that the results were mixed. The largest obstacles to CZE/XFX were a lack of reliable electricity, interference from other stations, and the popularity of commercial programming. Nevertheless, the station reached thousands of people. There were many stories of failure, but others of success. By the end of the Calles era, radio and state broadcasting were known entities in many parts of Mexico, and through radio, many residents became more connected with the state and the post-1929 state party, the PNR.

The biggest difficulty in establishing rural radio audiences was electricity. Many villages did not have it, and in communities that did, teachers—the usual recipients of the radios—did not always pay the electric bill. In other communities, an electric line existed and bills were paid, but service still proved irregular. In 1933, Professor Luis F. Rodríguez inspected seventy-one of the seventy-five communities where the SEP had donated relatively cheap, U.S.-made Atwater Kent receivers. He traveled through the states of Puebla, Tlaxcala, Hidalgo, and México. Out of the schools visited, twenty-one lacked electricity, five had not paid their electric bills, and one only obtained power after the owner of a small corn mill provided a current while the inspector was there. Radios could operate on batteries, side-stepping the electricity problem, but batteries did not last long, and it could take months before someone would obtain replacements in towns large enough to carry them.[38]

Another major problem for the SEP radio program was other stations. A large proportion of the schools in Puebla and Tlaxcala reported that they could only receive XFX during certain hours of the day, stating that the commercial station in Puebla—XIAJ—often overpowered the SEP signal. Some teachers and community members openly admitted that they enjoyed other channels, though many still listened to XFX on occasion or for a certain program. SEP officials themselves tuned in different channels for villagers. About his time in the small town of Acatlán, Puebla, inspector C. E. Sansalvador wrote that during the inauguration of the radio he played programs broadcast from American stations in San Antonio and Chicago because he had difficulty picking up the SEP's "bulletins of the revolution."[39]

Yet scholars have curiously failed to acknowledge the success stories. A number of teachers in the state of México said that they used the XFX singing program in their classes. Rodríguez reported that in Puebla, the school in the community of Moyotzingo was "without a doubt the school where they make the best use of the radio." He continued that over two hundred locals regularly gathered to listen to XFX, using

FIG. 22. Man with an Atwater Kent radio, c. 1929. Courtesy of Library of Congress, Prints and Photographs Division, Reproduction No. LC-USZ62-22205.

the programs to better their agrarian skills. All of the villages with radios in Tlaxcala reported listening to XFX in the mornings before the Puebla station took over.[40] In each community where the radios worked—about half—hundreds or thousands of people, depending on the location, came into contact with the devices and the broadcasts they trumpeted. When they worked, these radios were used. It cannot be accurately said that listeners always listened to the SEP station, but neither was it completely ignored.

All in all, the SEP brought thousands of rural people into contact with broadcasting. Despite the many ongoing problems of creating a state-radio listenership in the countryside, the SEP built a closer relationship between the federal government and a number of communities. In this regard, radio was only one of a number of important tools cultural missionaries applied in their endeavors. Movie projectors, farm equipment, cooking utensils, and a number of other objects also help SEP employees engage the hearts and minds of campesinos. But villagers especially prized the radios as a source of education, entertainment, and modernity. The SEP focused more on the central region of Mexico, but states including Guerrero, Nayarit, Oaxaca, Durango, Nuevo León, and even the Territory of Quintana Roo, also received SEP radios.

SEP Radio Beyond the Schools

It would paint an inaccurate picture of SEP broadcasting to discuss only the work done with schools. In addition to its educational endeavors, CZE, and then XFX, became the government's most successful medium for communicating political events to the nation and to foreigners abroad, at least until the PNR station became well established in 1931. The SEP itself was political. Its members eschewed speeches on secularizing education and building a nationwide allegiance based on narratives espoused by socialist and communist members of the revolutionary leadership.

Beginning with its first broadcast, the SEP station transmitted programs that directly relayed political messages and events. Not only did the station air presidential inaugurations, but also presidential addresses to Congress. In September 1929, XFX's coverage of President Emilio Portes Gil's remarks to Congress included not only the state of the union, but also the ceremonial events before and after that commemorated Mexican independence and the nation's heroes. Employees placed microphones for the station in the towers of the national cathedral in the zócalo (capital square), at the door of the entrance of the Chamber of Deputies, and within the congressional building itself. XFX broadcasted the military marches that accompanied the president along his brief trip from the executive offices to the legislative house. This event was immediately followed by a transmission of the traditional *grito de independencia* (Cry of Independence), which the station picked up with microphones in the Plaza de Armas and on the balcony of the National Palace. According to María Luisa Ross, the influential manager of the SEP station, the program was well received across the country and in parts of the United States. Residents from a multitude of states, including Puebla, Veracruz, Hidalgo, Guanajuato, Morelos, and Campeche sent in letters. She received similar correspondences from listeners in the Federal District. A man in Brownsville, Texas, wrote that he had even heard the bells of the cathedral with perfect clarity.[41] The tradition of broadcasting Independence would continue for decades as a prominent component of radio broadcasting and television.[42]

The SEP station also broadcast sessions of Congress. In her monthly report for September 1929, Ross stated that it had been regular practice to transmit from the Congress in the afternoon.[43] The previous year, when a correspondent for the *New York Times* published an article on what Mexican stations were most likely to be heard in the United States, he stated that CZE, despite its relatively weak strength, could be heard in "Southern and Central United States and in some parts of Central Canada." In addition to providing a typical airplay list of

CZE—cooking, exercise, domestic science, music, bedtime stories—it also discussed the wide variety of political programming on the stations, including "the Mexican Congress, Cámara de . . . the night of the 'Grito,' Sept. 15, which is the most important national celebration and when the President is present and takes part, the Mexican Independence ceremonies are broadcast by remote control. Also, remote control is extended to the Presidential palace at times for the special use of the president."[44] CZE/XFX, in other words, was the outlet for political propaganda.

The station also played an important role in attempting to diffuse public fears. The best example is the assassination of once-again president-elect Álvaro Obregón on July 17, 1928. For the first time, XFX and a number of other stations interrupted their broadcasts, providing news of the caudillo's death.[45] XFX continued by airing updates on political decisions. It also broadcast the trial of the assassin, José de León Toral, with the goal of showing that justice was served and to calm the public. Station employees placed a microphone before the judge's table. This backfired, however. Days later court authorities ordered the microphones removed because the defense "had taken advantage of this trial to spread seditious propaganda and to aid the clerical rebellion."[46] Certain SEP officials used the trial, along with the SEP broadcasts of war bulletins, to justify a proposal for building a more powerful government broadcasting station to reduce "fanaticism."[47] The proposal floundered, however, after the creation of the more powerful station built by the PNR in 1931.

Calles used his radio address during his annual speech to the Congress in September 1928 to declare the restructuring of the executive office and that the era of caudillos had ended. In its place would rise a new era of institutional politics, that of the single party.[48] Before the end of the month, Calles and his supporters, in collaboration with former allies of Obregón, chose Emilio Portes Gil as a compromise decision on the provisional presidency until new elections would be held in 1930.

The SEP radio station was far and away the most important government station during the 1920s, and it continued to play a vital role in the early 1930s. Calles and SEP bureaucrats used broadcasting to influence rural farmers, housewives, and children, and also as the direct mouthpiece for government proclamations, speeches, and events. It was not, however, the only government station tasked with building a popular base for the revolutionary state.

Radio and Labor

Labor organizations, like agrarian groups, constituted another arm of the ascending populist coalition. It is no surprise then that government leaders also focused radio programming on this growing portion of the population. Of course, the incorporation of workers into the political system predates the Calles presidency and, for that matter, broadcasting; it was a direct result of the Revolution itself. Leaders of every faction had espoused some form of populist rhetoric in decrees, newspapers, and public speeches. But broadcasting provided a much greater platform for corporatist and populist politics.

The Obregón administration began the relationship between broadcasting and labor. The president had made labor leader Luis N. Morones head of the National Military Manufacturing and Establishments. His department not only operated station VPD, but also expanded its radio operations among different factories. Under Morones's direction in 1923, the National Factory of Clothing and Equipment acquired broadcasting station 1-R and radiotelegraph station 1-Z.[49] As the head of CROM and a legislator during the early 1920s, Morones had a history of working with communications employees, and he enthusiastically took to radio. He had previous organized telephone and telegraph operators. By November 1923, the CROM had distributed a number of radios to workers in Mexico City, DF; Pachuca, Hidalgo; Puebla, Puebla; Orizaba, Veracruz; Torreón, Chihuahua; Jalisco; and "other places in the Republic."[50]

During the Calles years, radio listenership grew significantly. Under Luis N. Morones's leadership the SICT started its own "cultural" station

FIG. 23. Labor leader Luis N. Morones, 1935. Courtesy of Library of Congress, Prints and Photographs Division, photograph by Harris & Ewing, Reproduction No., LC-DIG-hec-39021.

under the call letters CZI in 1927. Two years later, after Morones's resignation, the station became XEFI, following new international protocols. According to historian Rosalía Velázquez Estrada, "all union groups affiliated with the CROM possessed a receiver and radio horn speaker in their gathering places."[51] The scant available data, however, fail to confirm this statement. What is known is that having this equipment was the goal, and that state and labor organizers had made significant headway by 1934. The U.S. Department of Commerce, for example, made special note that in 1930 the SICT had been placing radios among various labor organizations in Mexico.[52] And, as mentioned in chapter 4, the gifting and selling of radios to similar groups had been under way since late 1923.

The SEP also donated a large number of radios to labor groups in Tamaulipas in February 1930, the same month that Portes Gil turned the reigns of the presidency over to Pascual Ortiz Rubio. In fact, the order to ship the devices occurred the day before Portes Gil left office. This order is interesting not only because it occurred upon Portes Gil's departure from the presidency, but also because his greatest political support came from Tamaulipas, his home state. In that month alone, SEP officials, on behalf of Portes Gil, sent radio receivers to the Frontier Socialist Party, the League of Agrarian Communities, the Dockworkers Guild, the Campesinos Syndicate, and the Workers of Cía. México de Petróleo "El Aguila" Union—all of them in Tamaulipas. They also sent another device to a school in the state's capital Ciudad Victoria.[53]

Beyond the tie to Portes Gil, reasons for this sudden shipment of radios to Tamaulipas organizations are unclear. Relations between the government and foreign oil interests in Tamaulipas had been tense during the Calles years. Morones, Calles, and the Congress had pushed ahead stiffer laws further restricting foreign landownership in Mexico in 1925 and 1926. At the same time, the CROM provided declarations to the press that El Aguila, the largest petroleum company in the country, was waging a fearless and infamous campaign against Mexican workers, provoking conflicts in Tampico, Puerto México (present-day Coatzacoalcos), and Minatitlán.[54] During Portes Gil's last year in office there had been a fight in Congress over a new labor code and a raise in taxes on foreign oil interests.[55] Building a stronger connection between political and labor organizations in a state with a large presence of foreign companies made sense, even if relations with foreign oil companies had eased some in the late 1920s and early 1930s.[56]

These radio shipments occurred shortly after the formation of the PNR, of which Portes Gil was a leader. The Frontier Socialist Party had originally been formed to promote Portes Gil as a candidate for gubernatorial office in Tamaulipas, but the party was quickly subsumed by the national PNR after 1929. The timing of these gifts, the year after the PNR's formation, displays the increased pressure to unify political parties under the newly emerging one-party system.

Whatever the reason for the initial gifting of the Tamaulipas radios, federal agencies were giving out receivers to a myriad of different labor, political, and campesino groups. In addition to the Tamaulipas organizations, the SEP also sent a radio to the House of the Socialist Party in Iguala, Guerrero. The Partido Rojo del Sur Veracruzano had also possessed a radio since before Calles took office. Veracruz governor Tejada also used radios for communication with labor and campesino groups in that state. The SEP provided eight radios to the National Commission of Highways.[57] Giving out radios was a common way of building allegiances.

What exactly did 1-R, CZI, and XEFI broadcast? In many regards, programming on the labor radio stations was similar to commercial programming with the exception that the labor stations had political propaganda and economic news between music segments. A program listed in *Excélsior*, October 26, 1926, shows that there was a variety of song styles: fox-trots, *canciones yucatecas* (Yucatecan songs)—some sung by Guty Cárdenas—tangos, *danzón* (a dance and music influenced by African and European rhythms and styles especially popular in Cuba and Veracruz), and *corridos* (Mexican folk ballads) including "El Ferrocarril" ("The Railroad").[58] The latter genre may have been slightly more common on CZI than on commercial stations such as CYL, CYB, and CYJ. But opera and classical compositions were common on CZI and XEFI, as with the other stations. The SICT also broadcast regular news on the economic market, especially commodity prices, crop reports, and programs on foreign demands.

Like XFX, XEFI also broadcast political speeches. Under the direction of station managers Carlos del Pozo and Jorge Peredo—the latter had previously operated an experimental radio operation in 1925—XEFI worked with other stations to air the message of President Pascual Ortiz Rubio (1930–32) during Pan-American Day. The speech, which promoted unity, peace, and respect for national sovereignty was heard in the far corners of Mexico and in many parts of the Western Hemisphere.[59]

Although government stations helped relay political messages across Mexico, labor and organized campesino groups surely faced some of the same electric and reception problems that plagued the SEP's radio program in rural schools. Electricity was more accessible in urban centers and factories, but not in rural communities. Little documentation exists about whether laborers and campesino groups actually listened to SICT. Surely urban workers sometimes tuned in to foreign and local commercial programming. It is known that groups listened to addresses from presidential candidates and presidents. Perhaps, at times, labor leadership enforced a greater adherence to state radio, but evidence is scant.

Regional leaders, such as Felipe Carrillo Puerto, radical governor of Yucatán, also used broadcasting to reach labor groups. As discussed in chapter 3, this southern state already possessed a history of radio experimentation. Carrillo Puerto used the medium extensively for his intertwined work with the Socialist Party of the Southeast and the Central Resistance League. In the Yucatán, the main political party and the government had already become one and the same by 1923. The governor and his supporters established a radio station near Mérida, CYY of the Grand Socialist Party from the Southeast. They also bought and distributed sixteen receivers to strategic locations throughout the state. Through these radios, party members broadcast educational lectures and music. On October 31, 1923, CYY aired music and a talk given by Professor Eligio Erosa Sierra in Maya from a restaurant attended by the governor and his family. Carrillo Puerto was assassinated the following year during the De la Huerta Rebellion, but the station continued to provide similar programming. In 1925, station broadcasters transmitted a program in honor of Carrillo Puerto for the reinauguration of the station under the new call letters of CCY. The Mexico City commercial station CYB-El Buen Tono rebroadcast the event "to all of the republic and beyond the frontiers." By the early 1930s the Mérida station was relaying PNR speeches and events, as well as commercial programming, from Mexico City stations.[60]

XE del Partido Nacional Revolucionario

In addition to the SEP and the SICT, the PNR became an important state presence on the airwaves. Just as the PNR incorporated a number of regional parties, its station, XEO (later XEFO, and XEUZ short wave), became the dominant political voice on radio. With the assistance of government and commercial stations, PNR messages resonated throughout the country.

Following Obregón's assassination, Calles, his supporters, and a number of prominent Obregonistas worked out an agreement to smooth over political complications. As a result, Calles gained his place as *"Jefe Máximo"* of the Revolution in exchange for compromising with Obregonistas on the selection of Emilio Portes Gil to fill the presidency until new elections in 1930. Another result of this process was that Calles and allied political leaders formed the PNR to further incorporate and institutionalize revolutionary factions. By 1929 distinguished party members were giving political speeches for the new agreed-upon candidate—Pascual Ortiz Rubio—over the radio. To accomplish its goals, the PNR, like government agencies and regional leaders, distributed receivers to various organizations.[61]

The formal inauguration of XE–Partido Nacional Revolucionario–XEO was on New Year's Day 1931, though the proprietors of *Excélsior* had allowed the party to use its station since June 1929.[62] As the official station of the PNR, XEO was the only specifically political station in operation during the 1930s. Congress had officially banned all oppositional radio in laws dating back to 1926. As a result, the station, and Mexican radio in general, reinforced the one-party system. In 1931, the station operated with 5,000 watts of power, but government officials elevated its strength to 50,000 watts two years later. After the rise of XEO, the SEP's station XFX still continued to broadcast political programming, working in partnership with the PNR. But starting in 1931, XEO became the flagship of party broadcasting.

Notable attempts at antigovernment radio programming did occur from 1924 to 1935, but all met with repression from state officials. General Arnulfo Gómez used radio during his brief revolt in late 1927. He hoped to obtain popular American support and to persuade U.S. leaders to withdraw recognition of Calles by broadcasting to various newspapers "from several of the border cities in Texas and Arizona."[63] Of course, he was not in Mexico. During another anti-Calles protest on November 7, 1931, a small group of three

> alleged Communists, pistols in hand, entered the [commercial] radio station XEW, 'the voice of Latin America' . . . tied up the operator José Piña and proceeded to broadcast insults to President Ortiz Rubio, Minister of War Calles and Ambassador Clark, as well as accusations that American imperialism was attempting to provoke war on the part of China and Russia against Japan . . . The incident occurred after the conclusion of a concert in memory of the noted Mexican musician, Carlos Menéndez, given under the auspices of the Department of Public Education.[64]

Another account provided in the communist paper *El Machete* states that members of the Communist Party took control of the station as a celebration of the anniversary of the Russian Revolution. According to this article, the rebel airwave invaders defended the Soviet Union, railed against U.S. imperialism, and argued that Calles's repressive regime was responsible for Mexico's misery.[65] Later, in January 1935, shortly after the inauguration of Lázaro Cárdenas, commercial station XXX, associated with the Mexico City newspaper *La Prensa*, broadcast a program by the "Three Wise Men" that criticized the politicization of the youth and the weak attendance of children at events celebrating Three Kings' Day.[66] Although rare, critics of the political order managed to occasionally get their voices on the air.

Conversely, almost all broadcast stations played programming that supported the revolutionary government. No antigovernment or anti-PNR operation ever aired long. Commercial-station owners did not always agree with government policies, but they generally worked in partnership with the state. They played a wide variety of "national" pieces, agreed to government supervision, and provided airtime for state programming. Some of the most powerful radio operations, such as the chain of stations being built by XEW's founder Emilio Azcárraga, worked closely, if reluctantly, with the PNR.

XEO worked with commercial entities, but it also collaborated with government stations, especially SEP station XFX. Before the commencement of XEO in 1931, the PNR had already turned to XFX to help propagate the party's messages. XFX collaborated with the leadership of the newly formed party in political campaigns, including the Nationalist Campaign that promoted the consumption of Mexican-made products and domestic arts. The station also broadcast PNR bulletins, which usually consisted of newspaper reports that showed the party in a positive light.[67] The PNR moved to present itself as the sole political representation of the Revolution, and XFX became an early ally in promoting its platform.

After the PNR created its own broadcasting station, the party's leadership continued to work closely with XFX on programming. For example, J. M. Puig Casauranc and María Luisa Ross communicated frequently with Senator Silvestre Guerrero, the secretary general of the PNR, "to immediately make an intense labor of communication between both stations." Ross stated that she enthusiastically supported the collaboration and scheduled meetings between the managers.[68]

The programming of XEO consisted of political messages between musical pieces, and, at times, commercials. The station's managers consistently demanded top-notch performers for their broadcasts. Regular XEO performers included famed composer Miguel Lerdo de Tejada, Pedro Vargas, later called the "Nightingale of the Americas," and Alfonso Esparza Oteo, perhaps the most famous fox-trot performer in Mexico City. Talks aimed at "the popular masses" included "Revolutionary Concepts" and "News of General Interest."[69]

The PNR's first broadcasts were in service of Ortiz Rubio's presidential campaign. PNR-affiliated congressmen, generals, and journalists gave special addresses. Journalist Joaquín Piña gave a speech titled "Ortiz Rubio, a President for All Mexicans."[70] They also included operettas, Yucatecan songs, piano pieces, and a variety of domestic and foreign numbers. Occasionally an actor would talk and do a bit of radio-theater.

Ortiz Rubio's first radio speeches as president were a testament to the growing influence of political broadcasting in Mexico. As was already a tradition since Calles was sworn into office six years before, Ortiz Rubio broadcast his inaugural address across the nation on state and commercial stations. The pronunciation carried much of the same populist rhetoric that his three predecessors had espoused. It was heard in thousands of communities in Mexico and by inquisitive listeners in the United States. But it was an event shortly thereafter that became more memorable. Following the presidential address, Daniel Flores, a twenty-three-year-old man, fired six bullets at the president and his family in their car. One of them struck Ortiz Rubio in the jaw after cutting through his wife's ear. Although the assassination attempt rattled the president, leading to timidity and health problems, his radio monologue to the nation nineteen days later was reported in the United States to have had "the double effect of reassuring the country [Mexico] as to his recovery from the bullet and to the constructive policy he plans to adopt." The same report stated that another reaction to the broadcast was the "calming today of foreign exchange. The quotation for the peso rose and stocks and securities also showed unusual movement compared with previous days."[71] Broadcasting, especially in times of anxiety, proved influential in calming or exciting nationals and foreigners.

Ortiz Rubio extended his approval of PNR radio by inaugurating the official XEO station with a New Year's address—"A Message of Best Wishes to the People of the Nation." Subsequent presidents built on this tradition. The principal stations of the country also retransmitted the event, exhibiting another development that would continue to grow over the next decades—the collaboration of the owners of increasingly monopolistic commercial radio station chains with the official single party of Mexico.[72] In addition to Ortiz Rubio, Secretary General of the PNR Silvestre Guerrero and other members of the party's National Executive Committee attended the event. Manuel Jasso, the PNR's secretary of propaganda and culture gave a speech that described "XE del PNR" as a means to keep in "daily and constant contact with the collectives that invigorate the body [of the PNR] and still others that constitute our nationality" providing party doctrine, government information, and music by party-affiliated musicians. XE del PNR, Jasso declared, would reach the "most remote places of the Republic and far beyond the borders."[73] His speech clearly defined the station as a tool to build mass allegiance to the state and the party, which were equated with one another. It was a multipronged approach that spread "political, cultural, and social" programming across the (trans)national soundscape.[74]

Ortiz Rubio and the PNR during his presidency used radio skillfully. According to Ross, the previously discussed 1931 Ortiz Rubio speech for Pan-American Day was one of the most popular broadcasts of the department's fiscal year. XFX received "innumerable" letters from within Mexico, the United States, and Canada in appreciation of the president's address.[75] The speech exemplifies how the Maximato presidents capitalized on the increased regularity of the broadcasting industry and the rising number of receiver sets in the country.

Radio as a political tool continued to grow during the campaigns and presidencies of Abelardo Rodríguez (1932–34) and Lázaro Cárdenas. Former XFX technical manager engineer Javier Stavoli, who would go on to play a crucial role in developing Mexican television, switched over to station XEFO. His transition to XEFO is one of the more important markers of the decline of the SEP station at the expense of party radio. In January 1934, President Rodríguez "broadcast an explanation of Mexico's new minimum wage law of four pesos a day and pleaded for public support to make it a universal application."[76] Rodríguez used

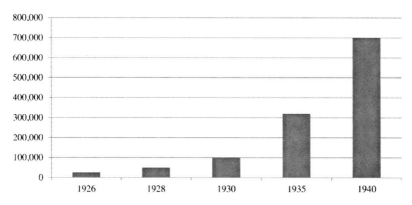

FIG. 24. Approximate number of radio receivers in Mexico, 1926–40. *Sources*: Hayes, *Radio Nation*, 33; figures adapted from the U.S. Commerce Department, *The Foreign Commerce and Navigation of the United States, 1926–35* (Washington DC: Government Printing Office, 1927–40).

the medium regularly, not only to popularize this policy, but also the image of the Mexican government in general. During his brief time in office, the collaboration between the PNR and commercial radiocasters expanded. When Rodríguez gave his last address to Congress in September 1934, XEFO had partnered with nineteen other stations to amplify the message. Many of the participating commercial broadcasters were located within Mexico City, like Azcárraga's XEW, but other stations—some affiliated with Azcárraga, some not—were located in Tamaulipas, Nuevo León, San Luis Potosí, Coahuila, Durango, Jalisco, and Veracruz.[77]

As a presidential candidate, Cárdenas spoke regularly over XEFO and its commercial and state broadcasting partners in Mexico City, Mérida, Tampico, Veracruz, Monterey, Saltillo, Nuevo Laredo, Querétaro, Torreón, Guadalajara, and San Luis Potosí. Even the border blaster XENT in Laredo, Nuevo León—owned by the famed gringo "cancer-curing" quack Norman Baker—transmitted PNR broadcasts of the Cárdenas campaign, providing another interesting dynamic in the relationship between these American broadcasters in Mexico's northern borderlands and the political leadership of the nation.[78]

Cárdenas "distributed hundreds of radios to his supporters in rural villages and working-class neighborhoods" during his fabled *gira* (campaign trek) across the country in 1934. Cárdenas also used a "radio train" to contact city officials and party members, among others, which, according to Joy Hayes, "signaled his commitment to the new medium."[79] The use of radio by the PNR had definitely increased. But as with the gifting of radios to labor groups and campesinos, Cárdenas's use of radio trains was not a new innovation. As discussed in chapters 3 and 5, rebel leaders and earlier presidents used radios on locomotives.

None of this detracts from Cárdenas's skill in political broadcasting. Cárdenas—along with Portes Gil who had acquired the land for the station—had been instrumental in the foundation of XEO as the president of the PNR in 1930.[80] He had become familiar with the medium's strengths and weaknesses well before he took office in late 1934. During his presidential campaign he used radio abundantly to further the reach of his call to increase social reforms and state cooperation with the common people. In one speech, "carried by all the radio stations in the Republic," he emphasized the rights of workers and said that a genuine corporatist system would reign supreme; he stressed the need to incorporate men and women and the importance of ending the exploitation of man by other men and by machines.[81] The oil expropriation speech he gave in 1938 electrified people from every nook and cranny of the country, elevating Mexican nationalism to an all-time high. Although some of this successful rallying can be attributed to Cárdenas's own intelligent use of the medium, the simple fact that stations expanded rapidly and abundantly during his presidency provided Cárdenas the essential tool. In other words, his ability to rally the populace stemmed directly from the maturation of radio technology and the Mexican broadcasting industry.

The Maturation of Radio

The first four years of the Cárdenas administration were tumultuous and exciting. The president exiled Calles and a number of his allies in 1936, nationalized the oil industry in 1938, fought a rebellion headed by San

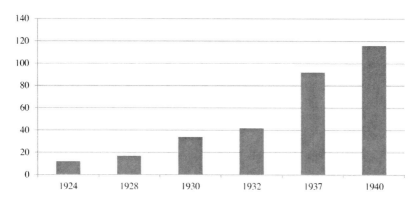

FIG. 25. Radio broadcasting stations in Mexico, 1924–40. *Sources*: "Estaciones trasmisoras de radio en la república," *Excélsior*, October 5, 1924, sec.2, 5; Sánchez Ruiz, *Orígenes de la radiofusión en México*, 38; Medina Ávila and Vargas Arana, *Nuestra es la voz*, 138, 156.

Luis Potosí strongman Saturnino Cedillo that broke immediately after the oil expropriation, and faced serious obstacles in implementing his radical land and labor reforms. But unlike problems encountered by national leaders during the 1920s, the Cárdenas administration faced much less risk of rebellion from within the military. The state had much more firmly secured its position. And though definitely not loved by all, Cárdenas had garnered more popular support than his revolution-ary predecessors. He not only appeared more genuine in his efforts to reach out to people, he had better means to do so. Subsequently, his words did not go unnoticed or without weight.

Radio use increased dramatically during the Cárdenas years. State and party broadcasting stations reached their pinnacle while massive commercial broadcasting chains solidified their dominant position on the airwaves. To maintain a strong presence in the face of the expanding private-sector operations, the government upped state messaging on commercial stations. Beginning in 1937, all large stations retransmitted the government's *The National Hour*.

By the early 1930s, radios had become common and loud enough in Mexico City that government health officials called for bans and limitations on the use of the devices in public. In 1931, SCOP agents had officially prohibited the common practice of blasting loudspeakers from storefronts, and city officials ordered saloons to turn off their radios by 9 pm. Not stopping there, they began enforcing noise ordinances against private residences where inhabitants played phonographs and radios too loudly.[82] According to certain members of the Department of Public Health, the noise caused serious health issues. They received a number of complaints. One woman told them that "radio noise had driven her mad." A Mexico City lawyer argued that his client died because radios prevented her from sleeping while ill.[83]

Beyond broadcasting to the public, state radio developed along lines established during the Calles era. The SCOP and Mexican military continued to use radio to suppress rebellions and to consolidate a stronger executive presence throughout the nation. Interestingly, much of the SCOP's efforts—and with the outbreak of World War II, the armed forces'—attention remained focused on the frontiers and coasts. The military, however, also increased the number of radio schools and devices in bases across much of the interior. Commercial and state broadcasters continued to broadcast abroad and to build closers ties with Central America. Emilio Azcárraga expanded his partnership with NBC in the United States, and Mexican stations continued to broadcast speeches from the president and other top party figures.

The armed forces under Cárdenas continued to professionalize and to increase its technological capabilities. In addition to importing tanks and new planes, the SGM established new radiotelegraph and radiotelephone offices and built new devices. During the fiscal year of 1935–36, the SGM reported that it had built sixty portable radios, two receptors, and two transmitters. The military sent and received just over 388,000 messages. The military established new transmission schools across the country.[84]

The SCOP continued to work on connecting central Mexico and the frontiers. Indeed, Secretary of SCOP Francisco J. Múgica stated in 1938 that the department was still giving special attention to linking the Baja California Peninsula and the Territory of Quintana Roo to Mexico City "in cooperation with the general plan of the government to incorporate these territories into national activities."[85] In early January 1936, it eliminated fees for official radio and wire communications between the Territories of Quintana Roo and Northern and Southern Baja California and authorities in other parts of the mainland. The ministry's employees also built 500-watt stations in Tuxtla Gutiérrez, Chiapas, and Oaxaca, Oaxaca. The department's workshop in Chapultepec was constructing two similar stations for Mérida, Yucatán, and Ensenada, Northern Territory of Baja California.[86] The following year they placed new Mexican-built transmitters for the stations in Guaymas, Hermosillo, La Paz, Mérida, Tapachula, and another for Coahuayana, Michoacán, "to keep the local public informed."[87]

The innovators of Mexican wireless continued to serve in both state and commercial communications positions. Azcárraga's broadcasting franchise continued to expand, including not only the chain affiliated with NBC, but another, headed by XEQ, affiliated with CBS. He maintained a working relationship with members of the ruling party, a sometimes rocky relationship that his descendants have continued to this day. José R. de la Herran, who cofounded SGM station JH and helped construct CYB and CYL in 1923, the Oaxacan Light and Power Company station in Oaxaca in 1924, and Monterrey's XET in 1930, installed and worked on XEW and XEQ's powerful transmitters from 1934 to 1940.[88] General Guillermo Garza Ramos y Trillo, who had worked with de la Herran in the private and government sectors, worked in a number of high-ranking positions in the Military Transmissions School during the Cárdenas years. He would serve as the General of Transmissions in the Baja California Peninsula during World War II starting in 1942. In 1955 he became the director of the military school he had worked for during the late 1930s. Meanwhile, he continued to install equipment for commercial stations, including XEW, XEQ, XEX, XETR, and XEN, among others.[89] Former Carrancista communications official and former president of the LCMR, Modesto C. Rolland, became Cárdenas's undersecretary of the SCOP. Felix Palavicini, the former Carrancista official, editor of *El Universal*, and CYL partner, founded XEN–Radio Mundial, a state-approved commercial station that provided international and national news in addition to popular and classical music. He would subsequently go on to become a XEW commentator in the 1940s.[90] Radio specialists tended to have long professional lives despite changes in administration and policy, and along with their careers, the radio industry they helped create was maturing.

Conclusion: Radio and the Making of Modern Mexico

It was during the Calles years, including the Maximato, that broadcasting became an essential component of state planning and propaganda. Although Carranza, and especially Obregón, had promoted their own strain of populist politics before—a necessary requirement for any leader of the Revolution—Calles brought populism to a whole new level, solidifying cross-class and corporatist alliances that brought larger numbers of people in contact with the Mexican state. Broadcasting was crucial to this endeavor. Government officials helped expand the number of radios provided to schools, labor groups, and government agencies, while middle-class and wealthy residents increasingly bought their own devices. Bars and other businesses also acquired their own receivers to play for patrons and employees. State stations, including the PNR operation, were not the favorite channels of most Mexicans, and efforts to get radio into rural communities faced numerous problems, not least of which was a lack of electricity. But even facing these seemingly unsurpassable barriers, radio reached larger and larger numbers of people in the late 1920s and 1930s. And many people did listen to the SEP and PNR stations, especially important political events and speeches. To make sure that government messaging was reaching public audiences, PNR legislators produced laws that forced commercial stations to air pro-government propaganda, and that restricted messages contrary to the desires of the state. Government officials and the owners of prominent stations also collaborated on the airing of music that was

"Mexican" in an attempt to build a stronger national culture, one that was more loyal to the revolutionary state and to Mexican commercial products. This partnership, and more importantly, the technology of radio broadcasting, allowed for this expansion of populist politics and nationalist sentiment, both of which Cárdenas built upon as he brought the political use of radio to new heights during the expulsion of Calles from Mexico in 1936 and the oil expropriation of 1938. Broadcasting had become a permanent part of government and modern Mexican culture.

Conclusion

Early Radio and Its Legacies

On the night of March 18, 1938, an important state message interrupted regular radio broadcasting. President Cárdenas declared that foreign petroleum companies had consistently and blatantly defied the laws of the nation and, for that, the state and workers were taking over much of the oil industry. Mexicans high and low celebrated the *grito* while U.S. and British oilmen, some of whom listened to the English translation of the speech, steamed in disbelief. The British recalled their diplomat and a number of American entrepreneurs clamored for swift retribution. Although Cárdenas, before and after, attempted to calm U.S. sentiments in other radio addresses, many residents of Mexico and the United States feared another rebellion or intervention.[1]

They got the former when longtime San Luis Potosí strongman General Saturnino Cedillo launched an insurrection just days after the oil declaration. He was upset over Cárdenas's agrarian policies and the president's efforts to weaken his grip on power in his home state of San Luis Potosí. Cedillo hoped to capitalize on animosities by persuading other generals, together with U.S. businessmen, to join him.[2] Like so many prominent rebels who had previously fought against the Mexican political order, he attempted to incorporate radio as a tool in his movement. In fact, he had been involved in a previous scandal involving radio messages and a plot to kidnap President Ortiz Rubio. At one point Cedillo managed to get a message broadcast from a U.S. border station in McAllen, Texas, just across the Rio Bravo from Reynosa,

FIG. 26. Mexican president Lázaro Cárdenas, 1937. Courtesy of Wikimedia Commons.

accusing Cárdenas of orchestrating a "communist dictatorship."[3] But within Mexico, federal operatives quickly discovered his wireless operations; worse for Cedillo, few people outside of a smattering of foreign businessmen and residents in San Luis Potosí genuinely supported the uprising. The army quickly put it down.[4]

In terms of wireless development, the oil expropriation and the Cedillo rebellion exhibited the complicated growth and evolution through which Mexican radio had passed in the preceding forty years. Cárdenas amped up his broadcasts on both the rebellion and the expropriation, appealing to a larger audience than preceding presidents and with greater means. Responses to Cárdenas's oil expropriation broadcast came from every region of Mexico. More than any previous point in Mexican history, it was clear that a professional army dominated military radio and that party leaders and partnering businessmen controlled broadcasting. These trends stemmed directly from the policies of Porfirio Díaz and the actions of revolutionary leaders who preceded Cárdenas.

There were a number of often intertwined continuities in radio development from the Porfiriato to the Cárdenas era: a strong state presence, the desire for control, military modernization, the incorporation of electronic wireless communications for centralization schemes, the use of radio to increase nationalism, and the advancement of communications with foreign powers. Connecting the frontier territories to the center of the nation remained an especially important goal. German specialists, SCOP employees, and military officers had built the first radiotelegraph towers in the Baja California Peninsula and the Territory of Quintana Roo. Although briefly interrupted by the Revolution, especially after the assassination of Madero, every single Revolution-era president, from Madero to Cárdenas, expanded wireless endeavors in the fringe territories and further increased Mexico's radio capabilities in Central America. Carranza used World War I as a means to gather more powerful equipment from Germany, fueling unstable relations with the United States, but greatly expanding his administration's reach via electronic communications. When World War II broke out in 1939, fears of German spies and poorly defended frontiers continued to worry the presses and governments of both Mexico and the United States. By this time, however, the Mexican state was more firmly behind the United States, and Mexico's communications system was more advanced, even if it paled in comparison to that of the Northern Colossus. Still, the drive for national unification and state expansion and the use of radio to accomplish these goals were ongoing processes.

Although these continuities are clear, the importance of radio, specifically on the outcome of the Mexican Revolution and, in turn, its impact on wireless development in Mexico, should not be overlooked. Following the onset of violence in 1910, the technology had become an important component of rebel and state communications, increasing correspondence capabilities between leaders and with foreign businessmen and government officials. Radio also became an essential tool in military intelligence gathering. Messages sent and overheard changed the course of some of the Revolution's most significant battles. All sides of the conflict used wireless, and it became an essential part of Constitutionalist designs to consolidate their power from 1915 to 1920. Venustiano Carranza and his cadre of advisers propagated new decrees and legislation that provided a stringent nationalist framework for radio, a response to the Revolution and U.S. military interventions. The laws provided a legal framework for confiscating illegally operating foreign stations and those of their domestic enemies. The radio regulations put in place were much more stringent than those during the Porfiriato, which were largely nonexistent excepting international agreements.

Despite attempts to control wireless, insurrectionists from Adolfo de la Huerta in 1924 to Saturnino Cedillo in 1938 found ways to use the technology to help organize their movements. However, the Mexican army, which increasingly became more professionalized during the 1920s, dominated the military use of radio following the De la Huerta Rebellion. Increased technological superiority together with a consolidating state hegemony, even if far from complete, helped presidents including Calles, Portes Gil, Ortiz Rubio, Rodríguez, and Cárdenas to put down armed challenges to their authority. Although divisions and conspiracies remained a common reality of political rule, radio became an important tool that helped maintain the governments under Sonoran revolutionaries Obregón and Calles and the single-party system.

Broadcasting became another essential component in building government support, amplifying state propaganda, and consolidating control, but it represents a technological departure from the preceding eras. State leaders and wealthy entrepreneurs understood the value of controlling the airwaves, especially in a society that had survived years of civil war and factionalism, and they compromised to establish a form of shared power. What better way to funnel a chaotic and popular revolution than with broadcasting—the world's greatest megaphone? More people were incorporated into the governing project, but voices of political dissent and religious proselytizing more often than not found their operations quashed. The revolutionary leaders in the national arena shaped these authoritarian tendencies, using force, favors, and legislation. They constructed their own broadcasting operations, but also turned to more successful commercial-station owners to carry their messages.

When broadcasting became a reality during the presidency of Obregón, the president opened up the medium to private development because of government weakness but also because of his capitalist leanings and his desire to make amends with commercial leaders who were interested in using the medium. While opening the field to private stations, he still put in place serious restrictions on what could be aired. He also initiated state broadcasting, which became a major component of radio from 1924 to 1938. Continuing rebellion forced state leaders to ban all radiocasts that attacked the government, increasing authoritarian trends in the name of stability and patriotism. The partnership between the government and the most prominent radio-station owners emerged as a key element of modern broadcasting in Mexico and reinforced the post-1929 one-party system and the monopolization of commercial broadcasting, trends that have continued to have a significant impact on Mexican society.

This consolidating and authoritarian strain in Mexican radio should not be surprising. The technology evolved in a period when revolution had fractured the state and when the victors raced to put it back together while incorporating larger bases of support. To figures such as Calles and Cárdenas, broadcasting had to be subjugated to the state's terms for matters of national security, national development, and political need, even while these presidents promoted the growth of the commercial sector. Even without revolutions, most forms of mass media have tended to start out relatively open and democratic, and then become more authoritarian and closed over time.[5] In the United States, for example, three corporations—the National Broadcasting Company, Columbia Broadcasting System, and the American Broadcasting Company—dominated radio broadcasting and television for decades. A handful of corporations still controlled the most powerful U.S. media networks as of 2016.[6] Radio proved especially prone to state and corporate control.

As journalist and technology specialist Timothy Wu wrote in 2011 about U.S. broadcasting, "radio becomes the clearest example of a technology that has grown into a feebler, rather than stronger, facilitator of public discourse."[7] This conclusion holds true for Mexico as well, albeit under different, and even more limiting, circumstances. After a violent revolution, powerful state and corporate transmitters quickly drowned out their amateur counterparts, which grew in number, but waned in influence. As most people contented themselves with owning a receiver to hear the news, advertisements, and entertainment programs, the proportion of listeners to transmitters dramatically increased. In comparison to their American counterparts to the north, or even their Argentine counterparts to the south, the democratic tendency—think of amateur or HAM operators communicating with each other within and across political boundaries—in radio had a smaller window in which to establish itself because it was interrupted by warfare and a revolutionary state bent on reconsolidating control.

The first forty years of radio development in Mexico displays similarities and differences with other nations and empires. Using wireless offices to connect frontier regions greatly mirrored how agents from Great Britain, France, and Germany relied on radio to build links with their own island possessions and far-flung colonies. The U.S. government similarly incorporated radio devices to place frontier territories under its political control, including Alaska and Hawaii. U.S. military officers and businessmen also established stations in the Philippines, Central America, Cuba, and Puerto Rico. Unlike these other powers, however, Mexico consistently relied on foreign specialists and technology. Its education and manufacturing institutions paled in comparison to the more industrialized nations. All of these countries also used wireless to advance their navies, of which Mexico had by far the smallest. But, in comparison to Central America and many other parts of Latin America, Mexico was economically powerful and technologically advanced. Despite its dependency on foreign technology, the Mexican government maintained a strong control over radio use in Mexico.

World War I caused all states involved to place radio under strict government supervision. In the United States, radio was largely a commercial and private affair, but as with other nations, the state greatly restricted, supervised, and largely took over wireless activities during the conflict. Even when the U.S. Congress returned radio to the hands of businesses and hobbyists, it worked to create new stringent codes, guidelines, and state oversight. The government also supported the formation of large private corporations that could informally propagate American political, as well as economical, interests. Britain and Germany subsequently made radio a state-directed medium. But a lack of serious studies about the effects of early wireless developments around the world still makes any sort of global comparison difficult.

Like its French, Austrian, Argentine, and Brazilian counterparts, Mexican broadcasting became a mixed system.[8] That is to say, the government promoted both state and private stations. Obregón's decision to allow commercial stations was influenced by a number of factors: he himself was an inventor and capitalist and supported such endeavors; a number of his most important advisers backed the idea; the government had little money to establish a successful radio network; private experimental operators were hard to stop and had advanced the field; and the state was still fragile and burdened with other problems. Another likely reason was the increasing presence of U.S. broadcasting and radio equipment. Receivers already started pouring southward over the lengthy U.S. border in the early 1920s, and Mexican listeners could pick up dozens of American stations. If the government was going to successfully establish its own cultural programs to counter U.S. programming, it would need the assistance, money, and connections of Mexico City and Monterrey capitalists. A similar argument has been made about the reasoning behind why the administration of Getúlio Vargas turned to commercial broadcasters in Brazil, except that in this case, the threat came from radio-savvy Argentina.[9]

Until the last year of Cárdenas's presidency, however, state agents kept a strong and direct presence in broadcasting. At its height, the government operated fourteen stations and often worked in partnership with the largest commercial broadcasters. Unlike Argentina and Brazil, Mexico had government broadcasting stations that not only operated, but that were influential in the 1920s, a result of Mexico's unique revolutionary circumstance and the correlating rise of populist and corporatist politics, which also came to Mexico earlier than in other parts of Latin America. Like state officials in France, Brazil, and Argentina, Mexican intellectuals and bureaucrats attempted to use radio to educate and bring "high culture" to the greater population. Many of these educators disliked the popular music programming on stations such as CYB or CYL, but some intellectuals and radio specialists were less critical; government broadcasting incorporated regional, popular, and even foreign music as much as the original commercial stations. Classical music dominated almost all stations in the 1920s, even if intermixed with *corridos, canciones yucatecas*, tangos, jazz, and fox-trots.

Yet by the 1930s, Emilio Azcárraga, with the assistance of NBC and CBS, came to monopolize more of the popular musicians, largely by providing better facilities and outpaying what the competition could offer. State stations relied on advertising at times, but it was never their main source of income, and the government, unlike in France, could not successfully implement a tax to fund the government stations. The Azcárraga family had proved a longtime and successful, if often contentious, partner of the government since 1923. Therefore, the PNR and its successors, the PRM and PRI, recognized the huge success and reach Emilio Azcárraga's operations had obtained and begrudgingly cooperated with him more often than they competed with him.

That partnership continues to the present day. Indeed, Emilio Azcárraga, his son Emilio Azcárraga Milmo, and grandson Emilio Azcárraga Jean became the most successful electronic media owners in Latin America, only recently surpassed in wealth and influence by Carlo Slim, who has come to dominate private telecommunications in Mexico. The Azcárraga family not only retained its prominent position in commercial radio during the 1940s, but also in television from the 1950s to the 1990s as owners of Telesistema Mexicano and then Televisa. They possessed approximately 80 percent of television viewership in Mexico and a similar percentage of advertising revenue, in addition to state funding.[10] The family did not always agree with the policies of incumbent presidents, whether it was Calles's war against the Cristeros or attempts to tax and regulate television, but they still publically remained allies of the party, especially after the 1940s. Azcárraga Milmo won notoriety for his blatant support of the PRI in the 1990s, stating that Televisa was a "part of the government system" and that he himself was the "number two *priista* [PRI member] in the country."[11]

Nonetheless, the media became more open in the late twentieth and early twenty-first century than it was in the mid-1900s. Following the

1985 earthquake, which killed over ten thousand people in Mexico, radio shows, reflecting an overall trend in Mexico City society, more openly chastised the government. Critical talk shows obtained larger audiences and, in turn, greater revenues. These stations forced government-aligned operations to rethink their strategies. Political reforms in the 1990s and the death of Emilio Azcárraga Jr. also opened up radio and media in general.[12] Still, complaints continue. The election of Enrique Peña Nieto in 2012 was mired by reports that Televisa, led by Azcárraga Jean, had unethically partnered with the candidate.[13] If true, it represents an ongoing partnership between the Azcárraga family and government officials that dates back to the 1920s.

The Mexican broadcasting industry originally sprang from the merger of Constitutionalist radio specialists with educated private experimenters and businessmen influenced by U.S. broadcasting trends. However, the state remained an important participant, mostly because of the continued (and not altogether erroneous) perception of radio as a security risk and because of the history of state-controlled radiotelegraphy operations in military matters and governance. A number of state officials also feared the domination of the airwaves by the country's northern radio provider, at least until a bilateral move toward accommodation between the United States and Mexico followed the oil nationalization, lingering financial difficulties, and the start of World War II.[14] Not all state involvement was shunned by Mexican station owners. Many of them embraced, if reluctantly at times, political participation, government collaboration, and state protectionism. Likewise, the official party favored the creation of vast commercial broadcasting chains, which greatly expanded the reach of government discourse.[15] As the single-party system became more firmly entrenched under the PRM and the PRI, and as Mexico rebuilt stronger partnerships with the United States during the Good Neighbor and World War II eras, the state reduced its direct broadcasting operations. But the partnership first constructed in the 1920s remained. Modern Mexico, born of revolution and radio, was under way.

NOTES

INTRODUCTION

1. "Desde la cúspide de una torre inalámbrica del bosque de Chapultepec se arrojó un sentimental y excéntrico suicida," *El Demócrata*, February 14, 1924, sec. 2, 1; "El hombre que cayó de la torre inalámbrica de Chapultepec no era un suicida sino un audaz agente de los rebeldes," *El Demócrata*, February 15, 1924, sec. 2, 1.

2. "Formal batida contra los aparatos radiotelefónicos," *El Universal Gráfico*, January 16, 1924, 2.

3. Coatsworth, *Growth against Development*, 35.

4. Craib, *Cartographic Mexico*, 127.

5. Coatsworth, *Growth against Development*, 36–37.

6. Noyola, *La raza de la hierba*, 19, 24–25, 56–57, 66.

7. J. C. Scott, *Seeing like a State*, 2.

8. Craib, *Cartographic Mexico*, 127–28.

9. Hayes, *Radio Nation*; Fernández Christlieb, *Los medios de difusión masiva en México*; Sánchez Ruiz, *Orígenes de la radiodifusión en México*; Mejía Barquera, *La industria de la radio y televisión*; Ortiz Garza, *La guerra de las ondas*; Ortiz Garza, *Radio entre dos reinos*; Medina Ávila and Vargas Arana, *Nuestra es la voz*.

10. Other publications that discuss prebroadcasting radio in Mexico are Ornelas Herrera, "Radio y cotidianidad en México"; Merchán Escalante, *Telecomunicaciones*; Fuentes, *La Radiodifusión*; Luz Ruelas, *México y Estados Unidos*.

11. For examples of works on communications and empires, see Innis, *Empire and Communications*; Headrick, *Invisible Weapon*; Satia, "War, Wireless, and Empire"; Yang, *Technology of Empire*; Wu, *Master Switch*.

12. For comparisons, see Karush, *Culture of Class*; McCann, *Hello, Hello Brazil*; McCann, "Carlos Lacerda."

13. Some of the general works on the Porfiriato and the Mexican Revolution that have influenced this work include, but are not limited to, Garner, *Porfirio*

Díaz; Bunker, *Creating Mexican Consumer Culture*; Krauze, *Mexico, Biography of Power*; Tannenbaum, *Mexican Agrarian Revolution*; Womack Jr., *Zapata and the Mexican Revolution*; Gilly, *The Mexican Revolution*; Knight, *Mexican Revolution*; Ruíz, *Great Rebellion*; J. M. Hart, *Empire and Revolution*; Katz, *Secret War in Mexico*; Katz, *Life and Times of Pancho Villa*; Ulloa, "La lucha armada (1911–1920)."

14. For examples, see Coatsworth, *Growth against Development*; Haber, *Industry and Underdevelopment*; Connolly, *El contratista de don Porfirio*; Garner, *Porfirio Díaz*; Agostoni, *Monuments of Progress*; Matthews, "*De Viaje.*"

15. Gallo's *Mexican Modernity* focuses on technology and culture.

16. See Popkin, *Revolutionary News*.

17. Thomas L. Friedman, "A Theory of Everything," *New York Times*, October 14, 2011, http://www.nytimes.com/2011/08/14/opinion/sunday/Friedman-a-theory -of-everyting-sort-of.html?src=ISMR_AP_LO_MST_FB, accessed August 14, 2011; Nathan Porter, "Social Media Stew Plays Potent Role in Global Digital Activism," *Washington Post*, October 13, 2013, http://www.washingtontimes.com /news/2013/oct/13/social-media-stew-plays-potent-role-in-global-digi/?page=1, accessed October 15, 2013; Howard and Hussain, *Democracy's Fourth Wave?*. For more recent research on digital activism, see http://digital-activism.org/projects.

18. Headrick, *Invisible Weapon*, 9. Also see Hong, *Wireless*, 92–93.

19. See Knight, *Mexican Revolution*, 2:148, 161. Knight also suggests that the first use of aerial propaganda occurred during the Mexican Revolution. See also Hagedorn, *Conquistadors of the Sky*, 76–77.

20. *Informes de las dependencias de la Secretaría de Comunicaciones y Obras Públicas del 11 de abril de 1919 al 31 de mayo de 1920*, 175.

21. Pertaining to the military influence on revolutionary politics, this work compliments Lieuwin, *Mexican Militarism*, 58; see also Navarro, *Political Intelligence*.

22. The PRI lost the presidency to the National Action Party (PAN) in 2000, officially ending the era of single-party rule. The PRI regained the presidency in 2012.

1. PORFIRIAN RADIO

1. "La telegrafía sin hilos," *El Mundo*, April 1, 1902, 1; Guzmán Cantú, "Telegrafía sin hilos," 11.

2. "La telegrafía sin hilos," *El Mundo*, February 15, 1902, 1; "La telegrafía sin hilos," *El Mundo*, April 1, 1902, 1.

3. *International Radio Telegraph Convention of Berlin, 1906*; "La telegrafía sin hilos," July 16, 1906, 2.

4. Garner, *Porfirio Díaz*, 114–45; Limantour, "El capital extranjero," 171; Grunstein Dickter, "¿Nacionalista porfiriano o 'científico extranjerista'?," 207–37; Bunker, *Creating Mexican Consumer Culture*, 1–11.

5. Aitken, *Syntony and Spark*, 233–35; Satia, "War, Wireless, and Empire," 848–50; Headrick, *Invisible Weapon*, 117–21.

6. Hong, *From Marconi's Black-Box to the Audion*, 41–42.

7. G. Marconi, "Origin and Development of Wireless," *New American Review* 168 (May 1899): 625, APS; "Ethereal Telegraphy," *Eclectic Magazine and Monthly Edition of the Living Age* [hereafter Living Age], December 3, 1898, 625, APS.

8. "Wonders of Wireless Technology," *Living Age*, October 16, 1897, 216, APS.

9. G. A. Esteva to the Secretaría de Estado del Despacho de Relaciones Exteriores, July 16, 1897, ASRE, caja 41–16–5.

10. G. A. Esteva to the Secretaría de Estado del Despacho de Relaciones Exteriores, August 13, 1897, ASRE, caja 41–16–5. Banti, a specialist in electricity, founded *L'elettricista* in 1892; Banti, *Il telefono senza fili sistema Marconi*.

11. G. A. Esteva to the Secretaría de Estado del Despacho de Relaciones Exteriores, April 12, 1899, ASRE, caja 41–16–5; M. Covarrubias to the Secretaría de Estado del Despacho de Relaciones Exteriores, May 10, 1901, ASRE, caja 41–16–5. Experimentation was also occurring in Russia, though Mexican political leaders did not interact with Russian officials as much as their more western European counterparts.

12. Adolfo Brule to the Señor Ministro de Relaciones Exteriores, November 9, 1898, ASRE, caja 41–16–5.

13. J. Beuif to Secretario de Relaciones Exteriores, June 1, 1900, ASRE, caja 41–16–5.

14. "Wireless Telegraphy for Africa," *Central African Times*, March 7, 1902, 4; "'Wireless' West Africa," *Sierra Leone Weekly News*, April 5, 1902, 7. For more on King Leopold II and the Congo, see Hochschild, *King Leopold's Ghost*.

15. The word "ethereal" was used frequently to describe radiotelegraphy in the 1890s and early 1900s. For example, "Ethereal Telegraphy," *Living Age* 69, no. 1 (January 1899): 41.

16. J. Beuif to Secretario de Relaciones Exteriores, June 1, 1900, ASRE, caja 41–16–5; J. Beuif to Secretario de Relaciones Exteriores, April 8, 1900, ASRE, caja 41–16–5.

17. Bernardo Reyes to the Secretario de Relaciones Exteriores, August 1, 1901, Mexico City, ASRE, caja 41–6–5; M. Covarrubias to the Secretario de Relaciones Exteriores, May 10, 1901, ASRE, caja 41–16–5.

18. [Illegible name] from the Mexican Embassy in the United States to the Secretario de Relaciones Exteriores, March 14, 1904, ASRE, caja 41–16–5.

19. Consulado de los Estados Unidos Mexicanos, Los Angeles, CA to Secretario de Relaciones Exteriores, April 26, 1905, ASRE, caja 41–16–5.

20. Francisco Z. Mena, *Memoria . . . por el Secretario Comunicaciones y Obras Públicas . . . 1 jul. 1900 a 30 jun. 1901* (Mexico City: Tipografía de Dirección General de Telégrafos, 1902), 169–70.

21. Martínez Miranda and de la Paz Ramos Lara, "Funciones de los ingenieros inspectores al comienzo de las obras del complejo hidroeléctrico de Necaxa," 231–86.
22. de los Reyes, *Medio siglo de cine mexicano*, 8–50.
23. Beezley, *Judas at the Jockey Club*, 16; Tenorio-Trillo, *Mexico at the World's Fairs*, 19, 37; Matthews, *"De Viaje,"* 251–89; Matthews, *Civilizing Machine*, 55–101.
24. Frank, *Posada's Broadsheets*, 187–91.
25. Frank, *Posada's Broadsheets*, 187–91.
26. Elihu Thomas, "Electrical Advances in the Past Ten Years," *Forum*, January 1898, 527–40, APS.
27. Wood, *Revolution in the Street*, 5.
28. "Wireless Telegraphy," *Los Angeles Times*, December 25, 1898, A5; John I. Waterbury, "The International Preliminary Conference to Formulate Regulation Governing Wireless Telegraphy," *North American Review* 177, no. 564 (November 1903): 656, APS; "Six Great Pioneers of Wireless," *EBU Technical Review* (Spring 1995): 90–92, http://tech.ebu.ch/docs/techreview/trev_263-pioneers .pdf, accessed June 24, 2011.
29. Mena, *Memoria, 1900–1901*, 169–70; Agustín M. Chávez, "Los directores de telecomunicaciones en la historia, *El Telegrafista* 3, no. 20 (April–May 1955): 8.
30. Merchán Escalante, *Telecomunicaciones*, 51.
31. Mena, *Memoria, 1901–1902*, 205.
32. Mena, *Memoria, 1901–1902*, 205; "La telegrafía y telefonía sin hilos," *Industria é Invenciones* (Barcelona), January 16, 1904, 8–9; "La telegrafía sin hilos," *El Mundo*, April 1, 1902, 1.
33. Mena, *Memoria, 1901–1902*, 205.
34. The value in U.S. dollars would be approximately half the number in pesos.
35. Bernardo Reyes, *Memoria . . . por el Secretario de Estado del Despacho de Guerra y Marina, 1 enero 1900 al 30 junio 1901*, tomo 1 (Mexico City: Tipografía de la Oficina Impresora de Estampillas, 1901), 323; "Continuaciones de 'el estudio,'" *Anales de Instituto Médico Nacional* (Mexico City: Oficina Tipográfica de la Secretaría de Fomento, 1894), 2; Ornelas Herrera, "Radio y contidianidad en México," 128. Ornelas Herrera argues that this radio experiment was the first successful one in Mexico.
36. Photo of the first radio experiment in Mexico, December 1900, AHI, Colección Fotografía Familiar. Although Díaz's sixth term in office began in December 1900, it would have marked his fifth reelection.
37. Noyola, *La raza de la hierba*, 19, 24–25, 56–57, 66.
38. Leandro Fernández, *Memoria . . . por el Secretario de Estado y del Despacho de Comunicaciones y Obras Publicas . . . 1 de julio de 1902 a 30 a junio de 1903* (Mexico City: Tipografía de la Dirección General de Telégrafos, 1904), 244.

39. "El Ministro de Comunicaciones," *El Mundo*, June 4, 1906, 1; Fernández, *Memoria, 1902–1903*, 244.

40. Secretaría de Fomento Colonización e Industria, *Censo de 1900* (Mexico City: Oficina Tipografía de la Secretaría de Fomento, 1901), A Hí, Colección Porfirio Díaz.

41. Taylor, "The Mining Boom in Baja California from 1850 to 1890 and the Emergence of Tijuana as a Border Town," 463–72; "En la Baja California se ha despertado gran interés," *El Economista Mexicano*, January 4, 1908, 207. For more information on El Boleo, see González Cruz, *La compañia El Boleo*.

42. Hurtado, "Empires, Frontiers, Filibusters, and Pioneers," 19, 37–43; Wyllys, "The Republic of California, 1853–54," 194–213; Rolle, "Futile Filibustering in Baja California," 159–66.

43. "Un pifia de 'The New York Herald,'" *El Dictamen* (Veracruz), March 18, 1908, 3.

44. "Un pifia de 'The New York Herald,'" 3; "Comunicación telegrafía con el Territorio de la Baja California, *El Imparcial*, September 20, 1910, 1.

45. Fernández, *Memoria, 1902–1903*, 245.

46. Fuentes, *La radiodifusión*, 24.

47. Fernández, *Memoria, 1902–1903*, 245; "El telégrafos sin hilos," *El Mundo*, December 9, 1902, 1.

48. "Telégrafos en California," *El Mundo*, October 1, 1903, 1; "Sonora," *El Mundo*, December 11, 1903, 2; "Sinaloa," *El Mundo, December* 11, 1903, 2; "Las vías de comunicación," *El Mundo*, August 19, 1903, 1; "Más líneas telegráficas," *El Mundo*, June 5, 1906, 1; "En la Baja California se ha despertado gran interes," *El Economista Mexicano*, January 4, 1908, 207.

49. "Más Líneas telegráficas," *El Mundo*, June 5, 1906, 1.

50. Fernández, *Memoria, 1903–1904*, 195.

51. Fernández, *Memoria, 1903–1904*, 195.

52. "Alemanes en Mazatlán," *El Mundo*, February 23, 1906, 2; Fernández, *Memoria, 1906–1907*, 102; "Telegrafía sin hilos," *El Mundo*, September 22, 1906, 1.

53. Fernández, *Memoria, 1908–1909*, 101.

54. de Dios Bonilla, *Apuntes para la historia de la marina nacional*, 244.

55. Fernández, *Memoria, 1908–1909*, 102.

56. Cleveland Moffett, "Marconi's Wireless Telegraph," *McClure's Magazine* 8, no. 2 (June 1899): 99–112.

57. See Sullivan, *Unfinished Conversations*; Rugeley, *Yucatán's Maya Peasantry and the Origins of the Caste War*; Rugeley, *Rebellion Now and Forever*.

58. British Honduras is present-day Belize.

59. Clegern, "British Honduras and the Pacification of Yucatan," 243–54; Macías Richard, *Nueva frontera mexicana*, 31–67.

60. Clegern, "British Honduras and the Pacification of Yucatan," 252.

61. *Convención entre los Estados Unidos Mexicanos y la Colonia de Honduras Británica para el enlace de sus líneas telegráficas*, 1910, Ramo Diplomático, ASRE, caja 7-14-58.

62. Fernández, *Memoria, 1906-1907*, 103-4.

63. Juan A. Hernández to Gral. de División Secretario de Guerra y Marina, December 17, 1910, AHSDN, Fondo Revolucionario, exp. xi/481.5/60, tomo 2; Juan A. Hernández to Gral. División Secretario De Guerra y Marina, December 18, 1910, AHSDN, Fondo Revolucionario, exp. xi/481.5/60, tomo 3; J. A. Hernández to Gral. Srio. Guerra y Marina, December 18, 1910, AHSDN, Fondo Revolucionario, exp. xi/481.5/60, tomo 3; "Another Victory for Mexican Rebels," *New York Times*, December 27, 1910, 4.

64. Merchán Escalante, *Telecomunicaciones*, 57.

65. Fernández, *Memoria, 1906-1907*, 103.

66. "La telegrafía sin hilos," *El Mundo*, January 2, 1906, 1.

67. Schwoch, *American Radio Industry and Its Latin American Activities*, 20.

68. "La telegrafía sin hilos," *El Mundo*, Janaury 2, 1906, 1.

69. "La entreviste de hoy: Guillermo Moreno Arenas," *El Telegrafista* 3, no. 24 (December 1955): 19-20.

70. The German ambassador to the U.S. Secretary of State, "International Wireless Telegraphy Convention," *Papers relating to the Foreign Relations of the United States with the Annual Message of the President, Transmitted to Congress, December 3, 1906*, pt. 2, 1513-14; Howeth, *History of Communications-Electronics in the United States Navy*, 547-48. For more on the 1903 and 1906 conferences, see Hills, *Struggle for Control*, 100-107.

71. *International Radio Telegraph Convention of Berlin, 1906*; "General Postal Union; October 9, 1874," Avalon Project: Documents in Law, History and Diplomacy, Yale Law School, Lillian Goldman Law Library, http://avalon.law .yale.edu/19th_century/usmu010.asp, accessed November 22, 2010.

72. Merchán Escalante, *Telecommunications*, 57.

73. *The Mexican Constitution of 1917 Compared with the Constitution of 1857*, 25. Article 28 established the government's right to monopolize postal services. In 1917, it would be applied to radio services as well.

74. Merchán Escalante, *Telecommunications*, 56-61.

75. "Aumento de empleado Telecátan," *El Diario*, July 8, 1907, 2.

76. "Aparatos de telegrafía sin hilos en el Golfo," *El Imparcial*, October 29, 1910, 5.

77. "La estación de telegrafía sin hilos Tres Marías," *El Imparcial*, October 10, 1910, 10; Merchán Escalante, *Telecomunicaciones*, 63.

78. Ornelas Herrera, "Radio y cotidianidad en México," 132-34.

79. Ornelas Herrera, "Radio y cotidianidad en México," 132.

80. Fernández, *Memoria, 1906-1907*, 102-03.

81. Ornelas Herrera, "Radio y cotidianidad en México," 132-33.

82. Fernández, *Memoria, 1909–1910*, 107; "La estación inalámbrica de Cabo Haro, destruida por el fuego, el sábado," *El Imparcial*, February 18, 1910, 1.

83. Bernardo Reyes, *Memoria, 1900–1901*, 36, 328, 359–60, 364–67; Manuel González Cosio, *Memoria . . . por el Secretario de Estado del Despacho de Guerra y Marina, 1 enero 1908 al 30 junio 1909* (Mexico City: Talleres del Departamento de Estado Mayor, 1909), 146.

84. "Telegrafía sin hilos," *El Mundo*, September 22, 1906, 1.

85. Fernández, *Memoria, 1906–1907*, 104; Fuentes, *La radiodifusión*, 24.

86. Fernández, *Memoria, 1906–1907*, 103–4.

87. Fernández, *Memoria, 1906–1907*, 104.

88. "Grandes utilidades que puede prestar a nuestro ejercito la telegrafía sin hilos," *El Imparcial*, April 6, 1910, 3.

2. RADIO IN REVOLUTION

Epigraph: Radiotelegrapher quoted in "Comunicaciones rápidas," *El Imparcial*, December 28, 1910, 5.

1. "Another Victory for Mexican Rebels," *New York Times*, December 27, 1910, 4; Juan A. Hernández to Gral. de División Secretario de Guerra y Marina, December 17, 1910, AHSDN, Fondo Revolucionario, exp. xi/481.5/60, tomo 2; Juan A. Hernández to Gral. División Secretario de Guerra y Marina, December 18, 1910, AHSDN, Fondo Revolucionario, exp. xi/481.5/60, tomo 3; J. A. Hernández to Gral. Srio. Guerra y Marina, December 18, 1910, AHSDN, Fondo Revolucionario, exp. xi/481.5/60, tomo 3.

2. There is still little consensus on the exact number of deaths and dispersed people directly related to the Mexican Revolution. Estimates range from one to over three million people. See McCaa, *Missing Millions*.

3. Knight, *Mexican Revolution*, 1:183–84.

4. Juan A. Hernández to Gral. de División Secretario de Guerra y Marina, December 17, 1910, AHSDN, Fondo Revolucionario, exp. xi/481.5/60, tomo 2; Juan A. Hernández to Gral. División Secretario De Guerra y Marina, Chihuahua to Mexico City, December 18, 1910, AHSDN, Fondo Revolucionario, exp. xi/481.5/60, tomo 3.

5. "Marconi's Telegraph," *New York Times*, January 23, 1898, 3.

6. J. A. Hernández to Gral. Srio. Guerra y Marina, December 18, 1910, AHSDN, Fondo Revolucionario, exp. xi/481.5/60, tomo 3; "La línea en Malpaso," *El Imparcial*, December 3, 1910, 1. TSH is an acronym for *telegrafía sin hilos or telefonía sin hilos*, which literally translates to telegraphy without wires and telephony without wires, or wireless.

7. "Sucesos del norte," *Diario Católico*, December 17, 1910, 1; "La telegrafía sin hilos," *El Imparcial*, December 28, 1910, 1, 5; "Los tropas federales llegaron a 'El Rosario' y se espera de un momento a otro encuentro las operaciones del ejército en el norte," *El País*, January 6, 1911, 1.

8. For more on Zapata, see Womack Jr., *Zapata and the Mexican Revolution*; Brunck, *¡Emiliano Zapata!*.

9. Stephan Bonsal, "Mexican Catholics Plan to Rule Nation," *New York Times*, May 23, 1911, 1.

10. Knight, *Mexican Revolution*, 2:201–5.

11. "El Sr. Gral. Díaz recibe noticia del temblor," *El Imparcial*, June 10, 1911, 1.

12. Ornelas Herrera, "La radiodifusión mexicana," 120–21; Ulloa, "La lucha armada (1911–1920)," 1102.

13. "Telegrafía inalámbrica en los cañoneros," *El Imparcial*, August 7, 1910, 9; "La telegrafía sin hilos en los barcos nacionales," *El Dictamen* (Veracruz), May 21, 1908, 1.

14. Merchán Escalante, *Telecomunicaciones*, 65; Fuentes, *La radiodifusión*, 25; "Agasajos á marinos del cañonero Bravo," *El Imparcial*, August 12, 1911, 5; David de la Fuente, *Memoria por el Secretario de Estado y del Despacho de Comunicaciones y Obras Públicas, 1911–1912* (Mexico City: Talleres Gráficos de la Secretaría de Comunicaciones y Obras Públicas, 1913), 129–30.

15. Fuentes, *La radiodifusión*, 25. Bonleper was the name of the French company that supplied the equipment.

16. Carlos Rincón Gallardo to the Ministerio de Gobernación, May 27, 1912, AHDF, Fondo Ayuntamiento, vol. 1254, exp. 400.

17. Works on Ricardo Flores Magón and the PLM invasion of Baja California in 1911 include Canovas, *Ricardo Flores Magón*; Sandos, *Rebellion in the Borderlands*; MacLachlan, *Anarchism and the Mexican Revolution*; Flores Magón, *Dreams of Freedom*.

18. "Vuelven á aparecer filibusteros en la Baja California," *El Imparcial*, July 19, 1911, 4; Cue Canovas, *Ricardo Flores Magón*, 50–53.

19. "Los magonistas merodean por Santa Rosalía," *El Imparcial*, July 24, 1911, 3.

20. "El Nuevo edificio de comunicaciones fue visitado por el Sr. Madero, *El Diario*, December 28, 1911, 1.

21. This finding reinforces similar conclusions reached by other scholars about the Mexican economy and industry from 1910 to 1913; see Haber, *Industry and Underdevelopment*, 124.

22. W. D. Macy to Henry Lane Wilson, November 11, 1911, NARA, RG 59, fold. 812.74/1.

23. Manuel Bonilla to the Secretary of Foreign Affairs, December 30, 1911, NARA, RG 59, fold. 812.74/1.

24. "Nuevas estaciones de telegrafía sin hilos," *El Imparcial*, June 2, 1911, 5; "Estaciones de telegrafía inalámbricas," *El Imparcial*, June 3, 1911, 3; "Telegrafía inalámbricas," *El Imparcial*, June 22, 1911, 5; "Estaciones de telegrafía inalámbricas en Campeche," *El Imparcial*, June 25, 1911, 4.

25. de la Fuente, *Memoria, 1911–1912*, 153; "El mensaje presidencial," *El Imparcial*, September 19, 1911, 5.
26. Merchán Escalante, *Telecomunicaciones*, 64.
27. de la Fuente, *Memoria, 1911–1912*, 129; Francisco I. Madero, "El Sr. Francisco I. Madero, al abrir las sesiones ordinarias del congreso, el 16 de Septiembre de 1912," in *Los presidentes de México ante la nación*, 2:38.
28. "Madero Killed, Havana Hears," *New York Times*, February 14, 1913, 2.
29. Victoriano Huerta, "El presidente Interino, Gral. Victoriano Huerta, al abrir las sesiones ordinarias del congress, el 1 de abril de 1913," *Los presidentes de México ante la nación*, 75. Although the unrest caused by the Madero rebellion prevented Mexico from sending a representative to the London wireless conference in 1912, the Madero administration did sign onto the agreement before the assassination of Madero in 1913. The U.S. government signed onto the international treaty as well.
30. Knight, *Mexican Revolution*, 2:71.
31. Henry Lane Wilson to Francisco León de la Barra, June 11, 1913, ASRE, caja 16-9-72; Francisco L. de la Barra to Henry Lane Wilson, June 12, 1913, ASRE, caja 16-9-72; Carlos Pereyra to the Secretario de Comunicaciones, June 13, 1913, ASRE, caja 16-9-72.
32. Peña y Reya to the Secretario de Comunicaciones, Mexico City, July 3, 1913, ASRE, caja 16-9-72.
33. Nelson O'Shaughnessy to Federico Gamboa, Mexico City, August 13, 1913, ASRE, caja 16-9-72.
34. F. Gamboa to Nelson O'Shaughnessy, Mexico City, August 15, 1913, ASRE, caja 16-9-72.
35. William Jennings Bryan to George Carothers, Washington DC, May 14, 1914, "Mexico," *Papers relating to the Foreign Relations of the United States, 1914* [hereafter cited as *FRUS*], 702, http://digicoll.library.wisc.edu/cgi-bin/FRUS/FRUS-idx?type=turn&id=FRUS.FRUS1914&entity=FRUS.FRUS1914.p0818&q1=Radio, accessed June 1, 2014.
36. "Bryan Calm on Huerta Note," *New York Times*, August 9, 1913, 2.
37. "Rebels Menace Line to Capital," *New York Times*, December 18, 1913; Gilly, *Mexican Revolution*, 91.
38. Merchán Escalante, *Telecomunicaciones*, 74.
39. Obregón, *Ocho mil kilómetros en campaña*, 130–31.
40. Cárdenas de la Peña, *Semblanza marítima del México independiente y revolucionario*, 2:224.
41. Quoted in Ornelas Herrera, "La radiodifusión mexicana," 133.
42. The *FRUS*, 1914–15, are full of radio messages from the Pacific Fleet.
43. "Huerta Gunboat to Guaymas," *New York Times*, June 21, 1914, 2.

44. "Huerta Gunboat to Guaymas," 2.
45. de Dios Bonilla, *Apuntes de la historia de la marina nacional*, 296.
46. "Huerta to Use Wireless," *New York Times*, January 18, 1914, 2; Victoriano Huerta, "El Gral. Victoriano Huerta, al abrir las sesiones ordinarias del congreso, el 1 de abril de 1914," *Los presidentes de México ante la nación*, 103–4.
47. General Subsecretario de Guerra y Marina to Secretario de Relaciones Exteriores, January 6, 1916, ASRE, caja 16-16-99. This box is filled with correspondence between the Carranza government and Harlé & Cie. over the remainder of a bill owed the French company for the radio equipment bought by Huerta.
48. Benbow, *Leading Them to the Promised Land*, 46.
49. Knight, *The Mexican Revolution*, 2:103–15.
50. Quoted in Ornelas Herrera, "La radiodifusión mexicana," 125.
51. Obregón, *Ocho mil kilómetros en campaña*, 116–17, 133.
52. Obregón, *Ocho mil kilómetros en campaña*, 126.
53. Knight, *Mexican Revolution*, 2:115.
54. "Say Federals Moved First," *New York Times*, March 16, 1914, 1; "Foes at Torreon Move to Battle," *New York Times*, March 16, 1914, 1.
55. "Say Villa Being Flanked," *New York Times*, March 26, 1914, 2; Knight, *Mexican Revolution*, 2:145–46.
56. Gonzales, *Mexican Revolution*, 14. For more on the convention, see Quirk, *Mexican Revolution*.
57. "Los torres inalámbricas situados en Chapultepec," *La Convención*, January 16, 1915, 8.
58. "El C. Presidente hizo una visita al palacio de comunicaciones," *La Convención*, January 4, 1915, 6; "Franquicias Telegráficas para los jefes militares," *La Convención*, December 21, 1914, 8.
59. "Carranza Forced to Give Up Capital," *New York Times*, July 20, 1915, 7; "Expects Carranza Reply This Week," *New York Times*, August 24, 1915, 6.
60. L. G. González to the Oficial Mayor Encargado del Cuartel General, March 1, 1915, AGN, Fondo Emiliano Zapata, caja 15, exp. 9.
61. L. G. González to the Oficial Mayor Encargado del Cuartel General, February 28, 1915, AGN, Fondo Emiliano Zapata, caja 15, exp. 9.
62. L. G. González to the Oficial Mayor Encargado del Cuartel General, March 1, 1915.
63. L. G. González to the Oficial Mayor Encargado del Cuartel General, March 1, 1915.
64. L. G. González to the Oficial Mayor Encargado del Cuartel General, March 3, 1915, AGN, Fondo Emiliano Zapata, caja 15, exp. 9; L. G. González to the Oficial Mayor Encargado del Cuartel General, March 9, 1915, AGN, Fondo Emiliano Zapata, caja 15, exp. 9.

65. L. G. González to the Oficial Mayor Encargado del Cuartel General, March 1, 1915; Illegible name to the Oficial Mayor Encargado del Cuartel General, Cuernavaca, March 16, 1915, AGN, Fondo Emiliano Zapata, caja 15, exp. 10.

66. L. G. González to the Oficial Mayor Encargado del Cuartel General, March 1, 1915.

67. L. G. González to the Oficial Mayor Encargado del Cuartel General, February 25, 1915, AGN, Fondo Emiliano Zapata, caja 15, exp. 9; L. G. González to the Oficial Mayor Encargado del Cuartel General, March 2, 1915, AGN, Fondo Emiliano Zapata, caja 15, exp. 9.

68. González, *Contra Villa*, 349.

69. Terrazas, *El verdadero Pancho Villa*, 180.

70. "Aliens Safe, Says Villa," *New York Times*, December 19, 1913, 2; Katz, *Life and Times of Pancho Villa*, 397–442.

71. Quirk, *Affair of Honor*, 8–31; Rosenberg, *Spreading the American Dream*, 64; Knight, *Mexican Revolution*, 2:68–71; Katz, *Secret War*, 167–84; Benbow, *Leading Them to the Promised Land*, 26–41.

3. REBUILDING A NATION AT WAR

Epigraph: Ignacio Bonillas, quoted in "Mexico Being Reconstructed," *New York Times*, November 5, 1916, 9.

1. Wood, *Revolution in the Street*, 22–23. Also see Ulloa, *Veracruz, capital de la nación*.

2. "Venustiano Carranza, al abrir las sesiones extraordinarias el Congreso, el 15 de abril de 1917," *Los presidentes de México ante la nación*, 157–58; "Una de las grandes obras de la revolución," *El Pueblo*, March 12, 1917, 10.

3. "Three Capitals Neutral," *New York Times*, December 23, 1914, 20; Peter Podell to Venustiano Carranza, January 8, 1915, CEHM, fondo XXI, carp. 24, fold. 2379.

4. Hart, *Empire and Revolution*, 310.

5. "Obregón pisa los talones al vil fierros," *El Dictamen*, August 4, 1915, 1.

6. "West Coast Wants Peace," *New York Times*, December 25, 1914, 6; Ornelas Herrera, "La radiodifusión mexicana," 155; "Está ya reconstruido el transporte de guerra 'Progreso,'" *El Pueblo*, January 10, 1917, 1; "Venustiano Carranza, al abrir las sesiones extraordinarias el Congreso, el 15 de abril de 1917," 157; Adolfo de la Huerta, in *Informes, abril 1919–mayo 1920*, 170.

7. "Expects Carranza Reply This Week," *New York Times*, August 24, 1915, 6.

8. Hart, *Empire and Revolution*, 310.

9. Pablo González to Venustiano Carranza, June 1, 1916, CEHM, fondo XXI, carp. 81, fold. 8985. 1; "La campaña del sur y la telegrafía inalámbricas," *El Pueblo*, February 21, 1917, 1.

10. John J. Pershing to General Frederick Funston, November 2, 1916, *FRUS*, "Mexico," 1916, 612–13.
11. "Interrupt Army Wireless," *New York Times*, March 31, 1916, 2.
12. "Interrupt Army Wireless," 2.
13. "Expect Carrancistas to Seize a Railroad," *New York Times*, June 27, 1916, 2.
14. Subsecretario de Hacienda to Venustiano Carranza, Mexico City, to Guadalajara, March 8, 1916, CEHM, vol. 480, 1.84.1.
15. Ignacio Bonillas, quoted in "Mexico Being Reconstructed," *New York Times*, November 5, 1916, 9.
16. Rolland, *A Reconstructive Policy in Mexico*, 7.
17. Richmond, *Venustiano Carranza's Nationalist Struggle*, 161–62.
18. "La campaña del sur y la telegrafía inalámbrica," *El Pueblo*, February 21, 1917, 1.
19. Knight, *Mexican Revolution*, 2:366.
20. Richmond, *Venustiano Carranza's Nationalist Struggle*, 162.
21. Luis Domínguez to Venustiano Carranza, March 30, 1917, AHSDN, caja 139, exp. xi/481.5/285.
22. Knight, *Mexican Revolution*, 2:375.
23. "Estaciones radiotelegraficas," *El Heraldo de México*, September 2, 1919, 6.
24. "Una de las grandes obras de la revolución," *El Pueblo*, March 12, 1917, 10.
25. "Mexico Plans Many Big Improvements," *Los Angeles Times*, July 7, 1917, 19.
26. "Optimism along Border," *New York Times*, March 30, 1916, 2.
27. "Que la reglamentada la instalación de estaciones," *El Economista*, October 26, 1916, 1.
28. *The Mexican Constitution of 1917 Compared with the Constitution of 1857*, 25.
29. Richmond, *Venustiano Carranza's Nationalist Struggle*, 110.
30. "Hear United States Will Embargo Food," *New York Times*, June 14, 1917, 11.
31. "Estación inalámbrica en Quintana Roo," *El Economista*, February 24, 1917, 3.
32. "Acaba de inaugurarse una importante estación radiográfica," *El Pueblo*, January 21, 1917, 6.
33. Rolland, *Informe sobre el Distrito Norte de la Baja California*, 38–39.
34. Rolland, *Informe sobre el Distrito Norte de la Baja California*, 51.
35. Katz, *Secret War in Mexico*, 416–17.
36. "Wireless Seized at Mexican Piers," *New York Times*, June 6, 1917, 6; "3 Noted Germans Interned for War in Hunt for Spies," *New York Times*, July 7, 1917, 1.
37. This German message transmitted in early 1917 proposed a Mexican alliance with Germany against the United States in exchange for the lands Mexico lost to the United States in 1846, if the United States joined the war against Germany.
38. Heinrich von Eckardt, quoted in Katz, *Secret War in Mexico*, 417.
39. Katz, *Secret War in Mexico*, 418–19.
40. Heinrich von Eckardt to Cándido Aguilar, January 15, 1917, ASRE, caja 17-9-278; Heinrich von Eckardt to Cándido Aguilar, January 31, 1917, ASRE, caja 17-9-278;

Cándido Aguilar to H. Von Eckardt, February 5, 1917, ASRE, caja 17-9-278; Cándido Aguilar to Alfonso M. Siller, February 12, 1917, ASRE, caja 17-9-278.

41. Cándido Aguilar to H. Von Eckardt, February 5, 1917, ASRE, caja 17-9-278.

42. "Llegaron a El Salvador los telegrafistas mexicanos," *El Pueblo*, January 8, 1917, 1; Fuentes, *La radiodifusión*, 26.

43. "Fraternidad salvadoreña," *El Pueblo*, January 3, 1917, 8.

44. Buchenau, *In the Shadow of the Giant*, 49–53, 108–9.

45. Buchenau, *In the Shadow of the Giant*, 121.

46. "El imperialismo de los Estados Unidos," *El Pueblo*, January 12, 1917, 3.

47. Venustiano Carranza, "Frente a frente," *El Pueblo*, January 12, 1917, 9.

48. As Manuel Estrada Cabrera is vividly portrayed in Asturias, *El señor presidente*.

49. Buchenau, *In the Shadow of the Giant*, 126.

50. "La estación de Chapultepec es un prestigio para México," *Revista de los Telégrafos Nacionales* 1, no. 1 (May 1921): 7–8; Katz, *Secret War in Mexico*, 421.

51. For more on the Zimmermann telegram, see Nickles, *Under the Wire*, 137–60, and Tuchman, *Zimmermann Telegram*.

52. Heinrich von Eckardt to Cándido Aguilar, January 31, 1917, ARSE, caja 17-9-278.

53. William Canada to the Secretary of State, March 10, 1917, NARA, RG 59, fold. 812.74/61; "La estación de Chapultepec es un prestigio para México," 8; Cárdenas de la Pena, *El telégrafo*, 125–26; "Se desean telegrafistas y celadores de telégrafos," *El Pueblo*, January 9, 1917, 2.

54. Fuentes, *La radiodifusión*, 25–26.

55. Katz, *Secret War in Mexico*, 420–23; Gustavo Reuthe, "Estudio sobre la radio-técnica: Su desarrollo en general y especialmente en la República," Mexico City, 1922, unpublished study, AGN, Fondo Secretaría de Comunicaciones y Transportes, caja 525, exp. 9.

56. "Don Venustiano Carranza, al abrir las sesiones ordinarias el Congreso, el 1 de septiembre de 1919," *Los presidentes de México ante la nación*, 356–57; "Contestación del Dip. Arturo Méndez, Presidente del Congreso," *Los presidentes de México ante la nación*, 381.

57. "100,000 Germans Are in Mexico," *New York Times*, March 1, 1917, 2; "Germany Seeks Alliance against US," *New York Times*, March 1, 1917, 1.

58. "100,000 Germans Are in Mexico," 2.

59. "100,000 Germans Are in Mexico," 2.

60. "Wireless from Mexico to Germany," *New York Times*, March 9, 1917, 1.

61. Katz, *Secret War in Mexico*, 419.

62. Katz, *Secret War in Mexico*, 420.

63. Frank R. McCoy, "Wireless Stations in Mexico—Memorandum for the Ambassador," March 12, 1917, NARA, RG 59, fold. 812.74/60; Henry P. Fletcher to the Secretary of State, NARA, RG 59, fold. 812.74/60.

64. "Wireless from Mexico to Germany," *New York Times*, March 9, 1917, 1.

65. Richmond, *Venustiano Carranza's Nationalist Struggle*, 149–56.

66. Buchenau, *Last Caudillo*, 88.

67. Juan Suárez, quoted in "Mexico to Break with Hun-Suarez, *Oklahoman*, June 20, 1918, 1.

68. Trinidad W. Flores, May 30, 1919, in Matute, *Contraespionaje político y sucesión presidencial*, 31, 60–62.

69. "Supplement of the Work and Activities of the Military Intelligence Division for the Week Ending October 3, 1918," p. 20, APECFT, Fondo Espías, exp. 030201, inv. 5, leg. 6/12.

70. Schuler, *Secret Wars and Secret Policies in the Americas*, 183–84. Schuler reveals a number of initiatives that the Carranza administration undertook to prepare for a possible war with the United States, including military and diplomatic alliances with Japan and Germany. But he overstates foreign policy concerns and underestimates how domestic issues influenced Constitutionalist actions. Carranza sought out communications equipment and armaments from Germany and Japan in order to defeat internal enemies more than to fight a war with the United States. Indeed, most of Carranza's war materials came from the United States, not Germany or Japan.

71. "Wireless Joins Mexican Capital and Germany," *New York Times*, March 9, 1917, 1.

72. Cárdenas de la Pena, *El telégrafo*, 125.

73. de la Huerta, in *Informes, abril 1919–mayo 1920*, 53, 170; "La estación de Chapultepec es un prestigio para México," 7; Reuthe, "Estudio sobre la radiotécnica: Su desarrollo en general y especialmente en la República," 10; Dr. Heller Krumm to the Secretario de Relaciones Exteriores, August 24, 1919, ARSE, caja 16-27-26.

74. M. Pérez Romero to Venustiano Carranza, September 18, 1918, ARSE, caja 16-27-28; Yang, *Technology of Empire*, 64–68; Richmond, *Venustiano Carranza's Nationalist Struggle*, 104, 159; "Carranza Seeks Japanese Favor," *New York Times*, March 10, 1917, 1; "Japan Sends Mexico Munition Machinery," *New York Times*, February 26, 1917, 1.

75. Knight, *Mexican Revolution*, 2:492; Hamnett, *A Concise History of Mexico*, 218.

76. Matute, *Contraespionaje político y sucesión presidencial*, 14.

77. Trinidad W. Flores, June 10, 1919, in Matute, *Contraespionaje político y sucesión presidencial*, 35.

78. Trinidad W. Flores, September 10, 1919, in Matute, *Contraespionaje político y sucesión presidencial*, 70.

79. Trinidad W. Flores, November 25, 1919, in Matute, *Contraespionaje político y sucesión presidencial*, 101–3.

80. Trinidad W. Flores, October 22, 1922, in Matute, *Contraespionaje político y sucesión presidencial*, 84.

81. Arthur Sears Henning, "Seek Carranza Loot," *New York Times*, May 13, 1920, 1.
82. de la Huerta, in *Informes, abril 1919–mayo 1920*, 53.

4. GROWTH AND INSECURITY

Epigraph: "Radio Exposition Given in Mexico," *Los Angeles Times*, December 16, 1923, sec. 2, 15.

1. Gonzales, "Imagining Mexico in 1921," 247–70; Delpar, "Mexican Culture, 1920–1945," 543–72.
2. "El contingente del ramo en la exposición comercial internacional del Palacio Legislativo," *Revista de los Telégrafos Nacionales* 1, no. 6 (October 1921): 4.
3. "El contingente del ramo en la exposición comercial internacional del Palacio Legislativo," 4–5; Gálvez Cancino, "Los felices del alba," 112–16; "Y llegó a comunicación sin cables: La primera transmisión de radiotelefonía en México," *Relatos e Historias de México* 35 (July 2011): 81–83.
4. Among others, Anda Gutiérrez, *Importancia de la radiodifusión en México*, 17; Figueroa Bermúdez, *¡Que onda con la radio!*, 41; Gálvez Cancino, "Los felices del alba," 109–23.
5. Ornelas Herrera, "Radio y cotidianidad en México," 145; Albarrán, *Seen and Heard*, 133.
6. Ornelas Herrera, "La radiodifusión mexicana," 184–93.
7. "Las primeras pruebas de la telefonía inalámbrica," *El Universal*, September 29, 1921, sec. 2.
8. "La República Mayor de Centroamérica será gobernado por un triunvirato," *El Universal*, September 30, 1921, 1; Medina Ávila and Vargas Arana, *Nuestra es la voz*, 63.
9. N. J. Quirk, "Wireless Telephony in the Navy," *Telephony* (January 1908): 30–33, http://earlyradiohistory.us/1908df1.htm, accessed March 30, 2012.
10. "Dos estaciones de telefonía inalámbrica ayer inauguradas en la Exposición Comercial Internacional," *Excélsior*, September 28, 1921, 1; "Telefonía inalámbrica en los faros," *El Universal*, October 7, 1921, 3; also see Ornelas Herrera, "La radiodifusión mexicana," 62–77.
11. "Adolfo de la Huerta, al abrir las sesiones ordinarias el congreso del 1 de septiembre de 1920," in *Los presidentes de México ante la nación*, 409.
12. *Informes, abril 1919–mayo 1920*, 175.
13. *Informes, abril 1919–mayo 1920*, 171.
14. Mora-Torres, *Making of the Mexican Border*, 183.
15. "La Fundidora Monterrey, S.A.," http://www.monterreyculturaindustrial.org /fundidora.htm, accessed February 26, 2010; Fernández Christlieb, *Los medios de difusión masiva en México*, 92–93.
16. Fernández Christlieb, *La radio mexicana*, 60; Figueroa Bermúdez, *¡Que onda con la radio!*, 41–42; Fernando Curiel, *¡Dispara Margot, dispara!*, 17–21. Although

it is possible that Tárnava transmitted entertainment programming in 1919, documentary evidence does not exist to back the claim. He never mentioned doing so in interviews with communication scholar Marvin Alisky; see note 17.

17. Alisky, "Educational Aspects of Broadcasting in Mexico," 26.

18. Mejía Barquera, "Historia mínima de la radio mexicana (1920–1996)."

19. "Carnet social," *La Revista de Yucatán* (Mérida), March 3, 1923, 4. PWX began service in 1922 and was picked up in many parts of Mexico.

20. Mejía Barquera, "Historia mínima de la radio Mexicana"; "De la vista a la más estación difusora radiotelefonía aprendimos que México rinde el debido culto al genial Marconi: Una interesante entrevista," *El Universal*, October 3, 1923, sec. 2, 7.

21. Velázquez Estrada, "La radiodifusión mexicana," 276; Mejía Barquera, "Historia mínima de la radio mexicana"; Ornelas Herrera, "Radio y cotidianidad en Mexico," 144.

22. Mejía Barquera, *La industria de la radio y televisión*, 35.

23. For information on Rolland's role in government, see de Fornaro, *Carranza and Mexico*, which has chapters by Enriquiz, Ferguson, and Rolland; Smith, "Carrancista Propaganda and the Print Media in the United States": 159–60; "A Decree Establishing Free Ports in Mexico," *Nation*, April 27, 1921, 632–33.

24. "Radio, J. M. Velasco y Cía.," *El Universal*, December 9, 1923, sec. 3, 4; "Una importante junta efectuada por los miembros de la Liga Central M. de Radio," *Excélsior*, July 22, 1923, sec. 3, 10; "Telegram Wireless Concert Heard in Mexico," *Salt Lake Telegram*, June 13, 1922, 13; Mejía Barquera, "Historia mínima de la radio mexicana."

25. Douglas, *Inventing American Broadcasting*, 300; Hayes, *Radio Nation*, 27–34; A January 1924 newspaper article lists over thirty businesses in the Asociación de Comerciantes de la Ciudad de México that sold American radio products in Mexico City. "La recepción por radio no está prohibida en México," *Excélsior*, January 27, 1924, sec. 2, 6.

26. "Radio Concert Honors Mexico," *Los Angeles Times*, September 17, 1922, sec. 2, 8; "Mexicans Now Fond of Radio," *Los Angeles Times*, August 21, 1923, 4.

27. Hayes, *Radio Nation*, 25.

28. Hayes, *Radio Nation*, 27.

29. For the first proposals to build private radio chains, see L. B. Rauthbaum, "Proyecto para la instalación de la radio-telefonía como medio para desarrollar la instrucción y cultura del pueblo mexicano," July 14, 1922, APECFT, exp. 74, inv. 4759, leg. 1; also see the proposals in AGN, Ramo Presidentes, Obregón-Calles Papers, caja 252, exp. 803-R-21; Mejía Barquera, *La industria de la radio y televisión*, 24–29.

30. Ingeniero en Jefe to the Ingeniero Consultar, November 17, 1922, AGN, Ramo Secretaría de Comunicaciones y Transportes, exp. 525/9.

31. "Estudio de las comunicaciones radiotelegráficas y radiotelefónicas," May 25, 1922, AGN, Ramo Secretaría de Comunicaciones y Transportes, exp. 525/9; Reuthe, "Su en general y especialmente en la República hasta su estado actual así como los urgentes problemas que implica en la práctica para el país," AGN, Ramo Secretaría de Comunicaciones y Transportes, exp. 525/9.

32. Antonio de Noriega and Leach, *Broadcasting in Mexico*, 15; Mejía Barquera, *La industria de la radio y televisión*, 36–37; Hayes, *Radio Nation*, 37.

33. Mejía Barquera, *La industria de la radio y televisión*, 36.

34. For many radio stations around the world during this period, 100 to 1,000 watts were fairly typical. By the 1930s, stations in Mexico blasted transmissions at over 100,000 watts. Although the higher the wattage, the more power the transmission, a 1,000-watt station in Mexico City could be heard across much of North America, Central America, and the Caribbean on a clear night. "Los permisos para las estaciones de radiotelefónia," *El Universal*, September 1, 1923, 3.

35. Mejía Barquera, *La industria de la radio y televisión*, 36; Modesto C. Rolland to Alvaro Obregón, May 9, 1923; Modesto C. Rolland to Alvaro Obregón, Mexico City, May 9, 1923, and June 5, 1923, AGN, Ramo Presidentes, Obregón-Calles Papers, caja 51, exp. 121-C-R-4; Alvaro Obregón to Modesto C. Rolland, June 8, 1923, AGN, Ramo Presidentes, Obregón-Calles Papers, caja 51, exp. 121-C-R-4; "Dio principio la Convención Nacional de Radio," *El Universal*, September 7, 1924, sec. 3, 5, 11; Medina Ávila and Vargas Arana, *Nuestra es la voz*, 80.

36. "Una interesante conferencia sobre radiotelefonía," *El Universal*, March 4, 1923, 4.

37. "Piden se anule el reglamento para el radio," *Excélsior*, June 14, 1923, sec. 2, 1.

38. Manuel Maples Arce, "T.S.H. (El Poema de la radiofonia)," *El Universal Ilustrado*, April 5, 1923, 19. The Stridentists were clever and often iconoclastic artists that celebrated modernity similar to Futurist poets in Europe, though with less fascist sentiments.

39. "Los artistas que tomaron parte en la inauguración, que anoche se efectuó, de la primera estación transmisora de radiotelefonia—'El Universal Ilustrado'—'La casa del radio,'" *El Universal*, May 9, 1923, sec. 2, 1; Rashkin, *Stridentist Movement*, 96.

40. Manuel Barajas, "Los filarmónicos y el radio," *Antena*, 1, no. 2 (August 1924): 12.

41. Fernádez Christlieb, *Los medios de difusión masiva*, 92–93.

42. "La feria del radio fue inaugurada por el Presidente de la República, ayer," *Excélsior*, June 17, 1923, sec. 2, 1; "La próxima feria de radio en la capital," *Excélsior*, May 27, 1923, sec. 3, 9.

43. Haber, *Industry and Underdevelopment*, 128–29.

44. "Una de las estaciones de radio en México," *Excélsior*, March 9, 1924, sec. 3, 7; "Estaciones trasmisoras de radio en la República," *Excélsior*, July 6, 1924, sec. 3, 11.

45. "Mexico Now Fond of Radio," *Los Angeles Times*, August 21, 1923, 4.

46. "Mexico Radio Station Opens," *Los Angeles Times*, October 19, 1923 sec. 2, 3.

47. "Horario de transmisiones," *Excélsior*, October 5, 1924, sec. 3, 11; "Horario de transmisiones," *Excélsior*, October 19, 1925, sec. 3, 8; "Horario de transmisiones," *Excélsior*, November 30, 1924, sec. 4, 9; "Estaciones transmisoras de Estados Unidos," *Excélsior*, October 26, 1924, sec. 4, 10.

48. "California Wares in Demand Abroad," *Los Angeles Times*, August 12, 1923, sec. 5, 13; "Mexico Now Fond of Radio," *Los Angeles Times*, August 21, 1923, 14; "Mexico Now Will Allow Broadcasting," *Los Angeles Times*, September 30, 1923, sec. 2, 5.

49. "Mexican Congress Begins Work Today," *New York Times*, June 21, 1920, 16.

50. "Mexican Radio Talks with Chile," *New York Times*, June 19, 1922, 3; "La estación de Chapultepec es un prestigio para México," *Revista de los Telégrafos Nacionales* 1, no. 1 (May 1921): 7–8; "American Wireless Centre to Be Here," *New York Times*, October 19, 1920, 13.

51. "Monroe Doctrine to Be Live Topic at Pan-American Conference," *New York Times*, March 4, 1923, 10.

52. Eduardo Ortiz to Álvaro Obregón, March 14, 1923, AGN, Ramo Presidentes Obregón-Calles Papers, exp. 223-C-4.

53. Eduardo Ortiz to Álvaro Obregón, April 16, 1923, AGN, Ramo Presidentes Obregón-Calles, exp. 223-C-4.

54. Buchenau, *In the Shadow of the Giant*, 153.

55. Baltazar Chávez to Alvaro Obregón, 22 Jul. 1923, AGN, Ramo Presidentes, Obregón-Calles Papers, Exp. 223-C-4.

56. Juan D. Bojórquez to Alvaro Obregón, June 15, 1923, AGN, Ramo Presidentes Obregón-Calles, exp. 223-C-4.

57. Subsecretario de Comunicaciones y Obras Públicas to Alvaro Obregón, Mexico City, June 16, 1923, AGN, Ramo Presidentes Obregón-Calles, exp. 223-C-4.

58. Sección Editorial, *La Gaceta* (San José, Costa Rica), September 18, 1923, AGN, Ramo Presidentes Obregón-Calles, exp. 223-C-4.

59. Buchenau, *In the Shadow of the Giant*, 153.

60. Sección Editorial, *La Gaceta*. Over the next couple years, the other Central American government added broadcasting equipment, but radio development in the region during this period is still lacking a good history.

61. "Una de la estaciones de radio en México," *Excélsior*, March 9, 1924, sec. 3, 7; "Nuevos testimonios del éxito de nuestra estación transmisora "Excélsior-Parker,'" *Excélsior*, April 13, 1924, sec. 3, 9; Plutarco Elías Calles, "Informe rendido por el C. General Plutarco Elías Calles, Presidente Constitucional de la República, en el primer año de su gobierno, ante la 31 legislatura, el 1 de septiembre de 1925," in *La educación pública en México*, 245.

62. Ad, "El Universal–CYL," *El Universal*, October 19, 1923, 7.

63. Schwoch, *American Radio Industry*, 21–22.

64. Sección Editorial, *La Gaceta*.

65. See the comment by U.S. delegate Allen H. Babcock in *Inter-American Committee on Electrical Communications, May 27–July 27, 1924*, 269.

66. Schwoch, *American Radio Industry*, 73.

67. Mejía Barquera, *La industria de la radio y la televisión*, 31–34.

68. "Rebels Say They Hold the Oil Fields," *New York Times*, February 17, 1924, 6.

69. Ornelas Herrera, "La radiodifusión mexicana," 213.

70. Haber, Razo, and Maurer, *Politics of Property Rights*, 68–72.

71. "Evacuation of Puebla Reported," *New York Times*, December 14, 1924, 2.

72. Although there are several works that specifically address the de la Huerta Rebellion, a thorough study of the conflict is still lacking. The best publications to date are Dulles, *Yesterday in Mexico*; Plasencia de la Parra, *Personajes y escenarios*; Castro Martínez, *Adolfo de la Huerta*.

73. P. A. del D. Zarate, "A los inspectores de división y radio en toda la red," December 8, 1923, APECFT, Archivo Plutarco Elías Calles, exp. 16, inv. 1341, leg. 3/4.

74. José Soto to El Jefe del Departamento Confidencial de esta Secretaría, January 16, 1924, AGN, Fondo Secretaría de Gobernación, caja 5, exp. 44.

75. José Soto to El Jefe del Departamento Confidencial de esta Secretaría, January 16, 1924.

76. El Jefe del Departamento to José Soto, January 14, 1924, AGN, Fondo Secretaría de Gobernación, caja 5, exp. 44.

77. José Soto to El Jefe del Departamento Confidencial de esta Secretaría, January 16, 1924.

78. "Formal batida contra los aparatos radiotelefónicos," *El Universal Gráfico*, January 16, 1924, 2.

79. "Formal batida contra los aparatos radiotelefónicos," 2; El acuerdo del General Gómez, *Excélsior*, January 16, 1924, 1; "Cual es la situación real de los receptores de radio," *El Universal Gráfico*, January 17, 1924, 3.

80. Gómez Chacón, "Carrillo Puerto y la radio en Yucatán," 190.

81. "Rebel Reinforcements Called Up," *New York Times*, December 20, 1923, 1; "Obregon Prepares Drive on Vera Cruz," *New York Times*, December 25, 1923, 5; "Un radiograma interceptado," *El Universal Gráfico*, January 18, 1924, 2. Figueroa had rebelled before de la Huerta, on November 30, 1923, but then afterward joined the larger rebellion.

82. "Rebel Reinforcements Called Up," *New York Times*, December 20, 1923, 1.

83. "Obregon Prepares Drive on Vera Cruz," *New York Times*, December 25, 1923, 5.

84. For more examples, see Ornelas Herrera, "La radiodifusión mexicana," 195–224.

85. Arturo M. Elías to Plutarco Elías Calles, February 4, 1924, APECFT, Archivo Plutarco Elías Calles, exp. 56, inv. 1379, leg. 10/10.

86. F. L. Pineda to Gral. de Div. P. E. Calles, March 8, 1924, APECFT, Archivo Plutarco Elías Calles, exp. 56, inv. 1379, leg. 10/10; "Fortín fue ocupado ayer

tras un sangriento combate, *El Demócrata*, February 5, 1924, 1924, 1; Rebels Predict Assault on the Capital," *New York Times*, February 22, 1924, 4; El General Serrano llego ayer procedente de Puerto Mexico, *El Demócrata*, April 4, 1924, 6; Quieren echar a pique los barcos, *El Demócrata*, April 13, 1924, 16; Obregon Regains Yucatan," *New York Times*, April 22, 1924, 11; "De La Huerta Still Claims Successes," *New York Times*, March 3, 1924, 3.

87. Dulles, *Yesterday in Mexico*, 260–61.

88. "Embarazado las operaciones," *El Universal Gráfico*, February 6, 1924, 1; F. Estévez to Luis G. Zepeda, January 25, 1924, APCEFT, Fondo Fernando Torreblanca, exp. 115, inv., 1521, leg. 1.

89. F. Estévez to Luis G. Zepeda, January 25. 1924; "Mensaje del Gral. Estrada interceptado," *Excélsior*, December 17, 1923, 5; Ornelas Herrera, "La radiodifusión mexicana," 196.

90. Bravo Izquierdo, *Un soldado del pueblo*, 155, 173, 208, 253; Ornelas Herrera, "La radiodifusión mexicana," 209–10.

91. "Many, many, many, loving memories." F. Estévez to Luis G. Zepeda, January 25, 1924, APECFT, Fondo Fernando Torreblanca, exp. 21, inv. 788, leg. 1; Fernando Torreblanca to Hortensia E. C. de Torreblanca, January 26, 1924, APECFT, Fondo Fernando Torreblanca, exp. 21, inv. 788, leg. 1.

92. "La tripulación completa de los barcos esta lista en Filadelfia y Nuevo Orleans," *El Demócrata*, February 5, 1924, 1; "El radio y la milicia," *El Demócrata*, March 7, 1924, 6; "Rebels Lose 116 in Pachuca Battle," *New York Times*, January 13, 1924, s8; Rebels Proclaim Tampico Blockade, *New York Times*, January 15, 1924, 8; "Warships Ordered Back to Veracruz," *New York Times*, January 31, 1924, 1.

93. "La radiografía utilizaren en operaciones militares," *Excélsior*, January 7, 1924. 1.

94. Medina Ávila and Vargas Arana, *Nuestra es la voz*, 104.

95. Ornelas Herrera, "La radiodifusión mexicana," 211, 213.

96. "Anoche se hizo verdadero arte frente al poderosos aparato transmisor radio-telefónico de 'el Buen Tono, S.A.,'" *El Demócrata*, March 9, 1924, 8.

97. "Querían echar a pique los barcos rebeldes," *El Demócrata*, April 13, 1924, 16; "Obregon Regains Yucatan," *New York Times*, April 22, 1924, 11. Interestingly and perhaps not surprisingly, after the rebels sabotaged the Mérida radio station, U.S. warships allowed "American interests" to use the navy's radios, especially the USS *Cleveland*.

98. "Las comunicaciones telegráficas y ferrocarrileras con Huajuapan, Oaxaca, quedaron reparadas en su totalidad," *El Demócrata*, March 29, 1924, 6.

99. "La interesante conferencia-concierto de hoy de la Liga Central Mexicana de Radio," *El Demócrata*, March 27, 1924, 2;

100. "El viernes será el mitin político del Cívico Progresista," *El Demócrata*, April 10, 1924, 1, 14.

101. Transcript of the Calles CYL radio address, APECFT, Archivo Plutarco Elías Calles, exp. 118, inv. 1583, leg. 1/4.

102. "El viernes será el mitin político del Cívico Progresista," *El Demócrata*, April 10, 1924, 1, 14; "El Gral. Calles envió anoche de viva voz a la nación un mensaje de alta transcendencia," *El Demócrata*, April 12, 1924, 1, 3–4.

103. Frank Bohn, "Calles and Anti-Calles in Mexico," April 7, 1924, *New York Times*, 16.

104. Plutarco Elías Calles, "To All the World," April 11, 1924, APECFT, Archivo Plutarco Elías Calles, exp. 215, inv. 456, leg. 1.

105. "Calles for Zapata Plan," *New York Times*, April 13, 1924, E1.

106. John Coryn, "Original Mexico Red Decides to Back Up Capital," *Chicago Daily Tribune*, May 15, 1924, 12.

107. "El Gral. Calles envió anoche de viva voz a la nación un mensaje de alta transcendencia," *El Demócrata*, April 12, 1924, 1.

108. Manuel Azamar to Gral. P. E. Calles, April 12, 1924, APECFT, Archivo Plutarco Elías Calles, exp. 215, inv. 456, leg. 1.

5. INVISIBLE HANDS

Epigraph: "Morrow Voices Plea Via Radio," *Los Angeles Times*, September 15, 1930, 7.

1. Lieuwen, *Mexican Militarism*, 86.

2. Joaquín Amaro, in Memoria por la Secretaría de Guerra y Marina, 1924–1925, 10.

3. Amaro, in *Memoria por la Secretaría de Guerra y Marina, 1930-1931*, 42.

4. Amaro, *Memoria por la Secretaría de Guerra y Marina, 1930-1931*, 42.

5. "Todos los cuerpos del ejército van a tener estaciones de radio," *Excélsior*, February 22, 1931, Fondo Joaquín Amaro, APECFT, exp. 255, inv. 566, leg. 1.

6. Abelardo Rodríguez, in *Memoria por la Secretaría de Guerra y Marina, 1931-1932*, 181.

7. Warren Johnson to Arturo M. Elías, July 30, 1927, APECFT, Fondo Presidentes, exp. 233, inv. 185, leg. 5/7; Amaro, in *Memoria por la Secretaría de Guerra y Marina, 1926-1927*, 16, 99; Amaro, in *Memoria por la Secretaría de Guerra y Marina, 1929-1930*, 131; Amaro, in *Memoria por la Secretaría de Guerra y Marina, 1930-1931*, 112.

8. Amaro, in *Memoria por la Secretaría de Guerra y Marina, 1929-1930*, 143.

9. "Dotación de aparatos de radio a los aeroplanos mexicanos," *El Cronista del Valle* (Brownsville TX), July 15, 1927, 1.

10. Estado Mayor Presidencial, "Informe de la Comisión Intersecretarial de Radio," May 1933, APECFT, Archivo Plutarco Elías Calles, exp. 95, inv. 1926, leg. 1; "Sesenta y siete escuelas de radio tiene el ejército," *El Universal*, January 9, 1929, APECFT, Fondo Joaquín Amaro, exp. 95, inv. 406, leg. 8/52.

11. Dulles, *Yesterday in Mexico*, 312; Buchenau, *Plutarco Elías Calles and the Mexican Revolution*, 124–26.

12. Dulles, *Yesterday in Mexico*, 312.

13. Bantjes, *As if Jesus Walked on Earth*, 62–25.

14. England, "The Curse of Huitzilopochtli," 230–32; Spicer, *Cycles of Conquest*, 83.

15. Amaro, in *Memoria por la Secretaría de Guerra y Marina, 1926–1927*, 89.

16. de la Pendraja, *Wars of Latin America*, 289; England, "The Curse of Huitzilopochtli," 232.

17. Amaro, in *Memoria por la Secretaría de Guerra y Marina, 1927–1928*, 16; Amaro, in *Memoria por la Secretaría de Guerra y Marina, 1926–1927*, 37.

18. Carelton Beals, "The Indian Who Sways Mexican Destiny," *New York Times Magazine*, December 7, 1930, 8–9, 19, APS.

19. de la Pendraja, *Wars of Latin America*, 295.

20. de la Pendraja, *Wars of Latin America*, 295; Medin, *El minimito presidencial*, 39–40, 50; Plasencia de la Perra, *Historia y organización*, 341–42.

21. Emilio Portes Gil, "Sobre la rebelión 'escobarista,'" in *Los presidentes de México ante la nación*, 5:698.

22. Fallaw, "Eulogio Ortiz," 143.

23. Julio Trens to Joaquín Amaro, September 18, 1929, APECFT, Fondo Joaquín Amaro, exp. 3, inv. 306, leg. 02/66.

24. "Por radio hacese propaganda en contra del gobierno," *Heraldo de México* (Los Angeles CA), September 11, 1926, 1; "El Gbno. trata de localizar las estaciones misteriosas de radio en México," *La Prensa* (San Antonio TX), September 22, 1926, 4.

25. Raúl Azcárraga Vidaurreta, interviewed by Felipe Gálvez Cancino, c. 1972. Transcript provided by Raúl Azcárraga Reyes Retana to José Luis Ortiz Garza in an interview, July 8, 1998, private collection of José Luis Ortiz Garza.

26. J. M. Puig Casauranc to Joaquín Amaro, December 17, 1925, APECFT, Fondo Joaquín Amaro, exp. 9, inv. 297, leg. 17/62.

27. Amaro, in *Memoria por la Secretaría de Guerra y Marina, 1927–1928*, 38.

28. Ramón Cortes González to the Secretario de Guerra y Marina, September 21, 1929, APECFT, Fondo Joaquín Amaro, exp. 18, inv. 248, leg. 1.

29. Amaro, in *Memoria por la Secretaría de Guerra y Marina, 1924–1925*, 10.

30. Amaro, in *Memoria por la Secretaría de Guerra y Marina, 1925–1926*, 66.

31. Amaro, in *Memoria por la Secretaría de Guerra y Marina, 1929–1930*, 30.

32. Estado Mayor Presidencial, "Informe de la Comisión Intersecretarial de Radio," May 1933, APECFT, Archivo Plutarco Elías Calles, exp. 95, inv. 1926, leg. 1.

33. The Mexican Music Company to the Secretario del Guerra y Marina, APECFT, Archivo Joaquín Amaro, February 21, 1928, exp. 1, inv. 286, leg. 58/63; The Mexican Music Company to Elisa L. de Amaro, February 21, 1928, APECFT, Archivo Joaquín Amaro, exp. 1, inv. 286, leg. 58/63.

34. Joaquín Amaro to E. Portes Gil, August 1, 1929, APECFT, Archivo Joaquín Amaro, exp. 2, inv. 290, leg. 1/3.

35. Loyo Camacho, *Joaquín Amaro*, 143.
36. José R. Campos, "Remite informe con la visita efectuada a la sección de radio-telegrafia del ejército brasileño," July 1928, APECFT, Archivo Joaquín Amaro, exp. 15, inv. 245, leg. 10/19.
37. Ezequiel Ruiz V., "Informes del Mayo de Infantería Ezequiel Ruiz V., comisionado en la República Argentina, para efectuar de radiotelegrafía, 1930–1931," APECFT, Archivo Joaquín Amaro, exp. 43, inv. 23, leg. 1.
38. Amaro, in *Memoria por la Secretaría de Guerra y Marina, 1930–1931*, 45, 88.
39. Andrés Figueroa, in *Memoria por la Secretaría de Guerra y Marina, 1934–1935*, 36–40, 64.
40. *Memoria por la Secretaría de Comunicaciones y Obras Públicas, 1927–1928*, 14; *Memoria por la Secretaría de Comunicaciones y Obras Públicas, 1935–1936*, 44.
41. "Importante servicio telegráfico," *Diario de Campeche*, August 23, 1926, 4.
42. Javier Sánchez Mejorada, in *Memoria por la Secretaría de Comunicaciones y Obras Públicas, 1928–1929*, 10; "Comunicación por radio con otras naciones," *El Nacional*, June 3, 1929, 5.
43. Cámara Nacional de la Industria de la Radio y Televisión, *Cápsulas culturales*, 250.
44. Note regarding the Montgomery Company, June 22, 1931, NARA, RG 59, 812.76/69.
45. Note regarding the Montgomery Company, June 22, 1931.
46. "Decreto por medio del cual se reglamenta el servicio radiotelegráfico internacional que se transmita por la vía 'Radiomex, R.C.A.," June 17, 1932, BMLT, Archivos Económicos; "New Wireless to Mexico," *New York Times*, January 7, 1932, 37. All Radiomex telegrams listed the locations, prices, etc., on the back of the telegrams. See, e.g., José M. Puig to Koki Hirota, Mexico City to Tokyo, October 24, 1934, ASRE, exp. III-297-14.
47. "Estación transmisora de radio en la ciudad de Chihuahua," *El Universal*, December 13, 1923, sec. 2, 7.
48. "Una amplia labor administriva desarrollada por el H. Ayuntamiento de Mérida, Yuc.," *Excélsior*, September 16, 1925, sección de rotograbado.
49. Gral. Calles to Gral. H. Jara, September 22, 1926, APECFT, Archivo Plutarco Elías Calles, exp. 11, inv. 2960, leg. 1.
50. Estado Mayor Presidencial, "Informe de la Comisión Intersecretarial de Radio," May 1933, APECFT, Archivo Plutarco Elías Calles, exp. 95, inv. 1926, leg. 1.
51. "Las cartas de rutas aéreas," *El Universal*, January 17, 1927, 1.
52. "Una ruta aérea entre la ciudad de México y Tapachula, Chiapas," *El Universal*, April 12, 1927, sec. 2, 1.
53. *Memoria por la Secretaría de Comunicaciones y Obras Públicas, 1927–1928*, 5.
54. Estado Mayor Presidencial, "Informe de la Comisión Intersecretarial de Radio," May 1933, APECFT, Archivo Plutarco Elías Calles, exp. 95, inv. 1926, leg. 1.

55. Emilio Portes Gil, quoted in "Mexican President Flies Over Volcano," *New York Times*, December 12, 1929, 14. The translation is from the original article.

56. Rebel Coup Triumphs," *Los Angeles Times*, December 4, 1931, 1.

57. "Servicio telegráfico a bordo de los ferrocarriles," *Excélsior*, February 12, 1928, 3.

58. "Grandes obras de irrigación que van a construirse en breve en Chihuahua," *Excélsior*, November 15, 1925, APECFT, Fondo Joaquín Amara Camaro, exp. 165, inv. 456, leg. 10/12.

59. "Mexico's Head Gets Fine Train," *Los Angeles Times*, May 15, 1927, 6.

60. *Memoria por la Secretaría de Comunicaciones y Obras Públicas, 1927-1928*, 10.

61. *International Radiotelegraph Conference of Washington, 1927.*

62. Secretaría de Comunicaciones y Obras Públicas, *Ley de Comunicaciones Eléctricas*; Mejía Barquera, *La industria de la radio*, 30–31; Hayes, *Radio Nation*, 37.

63. "Proposals Offered by the Mexican Delegation on the Subject of the Agenda," June 11, 1924, NARA, RG 59, box 574, D4/179.

64. "Proposals Offered by the Mexican Delegation."

65. Article 10 of the *Ley de Comunicaciones Eléctricas*, in Medina Ávila and Vargas Arana *Nuestra es la voz*, 83.

66. *International Radiotelegraph Convention of Washington, 1927*, 16–17, 35.

67. Ortiz Garza, *Una radio entre dos reinos*, 22.

68. See the telegrams between the governor of Coahuila to the Mexican Secretary of Communications, July 3–8, 1931, AGN, Ramo Presidentes Obregón-Calles, caja 69, exp. 8/4652.

69. While operating the station in Villa Acuña, Coahuila, Brinkley changed his focus to colon procedures.

70. Kahn, "The Carter Family on Border Radio," 205–17. For more works on border radio, see Robles, "Shaping *México Lindo*"; Rudel, *Hello, Everybody!*; Fowler and Crawford, *Border Radio*; Miller, *On the Border*.

71. "Maybe It's a Goatee," *Los Angeles Times*, January 2, 1932, A4. A Van Dyke beard, named after seventeenth-century painter Anthony van Dyke, consists of any growth of both a moustache and chin hair with the shaved cheeks. According to some nineteenth- and twentieth-century newspaper columnists, it indicated a selfish, sinister, or pompous man. See Peterkin, *One Thousand Beards*, 172–73.

72. "Fight Radio Station in Mexican Border," *New York Times*, December 28, 1931, 13.

73. Delpar, *Enormous Vogue of Things Mexican*; J. C. Delgadillo to the Secretaría de Comunicaciones y Obras Públicas, June 22, 1933, AGN Ramo Gobernación, caja 144, exp. 22/131.6-(721.1); Ortiz Garza, *Una radio entre dos reinos*, 49–87, 124.

74. See Ortiz Garza, *Una radio entre dos reinos*.

75. Medina Ávila and Vargas Arana *Nuestra es la voz*, 148–49; Ortiz Garza, *Una radio entre dos reinos*, 138, 144–47; Mariano Moctezuma to Julio Santoscoy, February 19, 1934, AGN, Ramo Gobernación, Caja 151, exp. 22/131.6 (721.3), 11; "Pen Points," *Los Angeles Times*, July 20, 1933, A4.

76. "Mexico to Discuss Radio Pact with US," *New York Times*, September 4, 1932, 15; "Named for Radio Parley," *New York Times*, February 7, 1933, 23.

77. Miller, *On the Border*, 78.

78. Daniels, *Shirt Sleeve Diplomat*, 373–75; Also see Lee, *Bizarre Careers of John R. Brinkley*, 169.

79. Robert D. Hienl, "Mexican Radio Parley Is Still in Deadlock," *Washington Post*, August 7, 1933, 7; C. B. Joliffe, in *Hearings on Remote Control Border Stations—H.R. 7800*, 14.

80. "Radio Conference Opens in Mexico," *New York Times*, July 11, 1933, 6; "Broadcasters Feel Secure," *New York Times*, August 20, 1933, X7; Harry Nicholas, "Channels Baffle Radio Conference," *New York Times*, August 13, 1933, B3; Robert D. Hienl, "Mexican Radio Parley Is Still in Deadlock," *Washington Post*, August 7, 1933, 7.

81. Fowler and Crawford, *Border Radio*, 331–32; Lornell, *Exploring American Folk Music*, 61.

82. For a discussion on how mass media tend to start open and democratic and then become more authoritarian and closed, see Wu, *Master Switch*, 5–39.

6. BROADCASTING STATE CULTURE

1. Velázquez Estrada, "La radiodifusión mexicana," 294.

2. Hayes, *Radio Nation*, 77.

3. Alvaro Obregón, "El Gral. Alvaro Obregón, al abrir las sesiones ordinarias el Congreso el 1 septiembre de 1924," *Los presidentes de México ante la nación*, 638.

4. Medina Ávila and Vargas Arana, *Nuestra es la voz*, 100.

5. Booth, *Mexico's School-Made Society*, 145; Fell, *José Vasconcelos*, 257.

6. "Las ideas fundamentales de la Secretaría de Educación Pública en materia educativa," *El Universal*, January 1, 1926, sec. 4, 3.

7. "Mexico Radios Culture," *Los Angeles Times*, March 7, 1926, F11.

8. "Una gran estación de radio de la Sria. de Educación," *El Universal*, June 21, 1924, sec. 2, 1, 8.

9. "Teach Indians with Radio," *Los Angeles Times*, November 22, 1925, 1.

10. "Mexico Radios Culture," *Los Angeles Times*, March 7, 1926, F11.

11. Hayes, "National Imaginings on the Air," 243.

12. "Radioconiertos," *Excélsior*, January 18, 1928, sec. 2, 3.

13. "Radioconiertos," *Excélsior*, January 18, 1928, sec. 2, 3; "Radioconciertos," *Excélsior*, June 25, 1928, sec. 2, 3; "Conciertos," *El Nacional*, June 15, 1929, 5.

14. Alejandro Michel, "Programa de acción de la obra de Extensión Educativa por Radio de la Secretaría de Educación Pública," July 2, 1930, AHSEP, caja 9478, exp. 4.

15. "La radio como vocero de la alegría," *El Universal*, February 6, 1926, 5.

16. C. E. Sansalvador, "Programa que se desarrolló en la escuela federal 'Emilio Carranza' con motivo de la inauguración de un aparato de radio," April 11, 1929, AHSEP, caja 9474, exp. 25.

17. Jacobo Dalevuelta, "El radio como vocero de la alegría del Carnaval," *El Universal*, February 7, 1926, 1.

18. Sánchez, *Mexico*, 85–86; Tovey, "The Role of the Music Educator in Mexico's Cultural Missions," 3–4; Torres, *Education and Social Change in Latin America*, 100.

19. Luis F. Rodríguez to the Jefe del Departamento de Enseñanza Rural y Primaria Foránea, March 18, 1923, AHSEP, caja 9482, exp. 78.

20. Albarrán, *Seen and Heard*, 129–41.

21. Alejandro Michel, "Informe de la labores desarrolladas por la obra de Extensión Educativa por Radio de la Secretaría de Educación Pública, durante el mes de marzo de 1930," March 1930, AHSEP, caja 9478, exp. 11.

22. Luis F. Rodríguez, "Informe de la inspección hecha en la zona de Texcoco, May 13, 1933, AHSEP, caja 9482, exp. 78.

23. Albarrán, *Seen and Heard*, 170–74.

24. "La obra cultural de la estación XFX de la Secretaría Educación Pública," *El Universal*, December 12, 1929, sec. 4, 9.

25. Hayes, "National Imaginings," 249. Also see "La obra cultural de la estación XFX de la Secretaría Educación Pública," *El Universal*, December 12, 1929, sec. 4, 9.

26. María Luisa Ross, "Estación C.Z.E.: Obra de extensión educativa por radio de la Secretaría de Educación Pública—programa," April 1927, FPECTFT, Fondo Joaquín Amara Camaro, exp. 9, inv. 297, leg. 15/62; Alejandro Michel, "Programa de acción de la obra de Extensión Educativa por Radio de la Secretaría de Educación Pública," July 2, 1930, AHSEP, caja 9478, exp. 4; Velázquez Estrada, "La radiodifusión mexicana," 177.

27. María Luisa Ross, "Informe de las labores desarrolladas por la Obra de Extensión Educativa por Radio de la Secretaría de Educación Pública, durante el mes de marzo de 1931," March 1931, AHSEP, caja 9478, exp. 4.

28. Moisés Sáenz, "Las escuelas rurales y el progreso del indio," *Mexican Folkways* 4, no. 1 (1928): 74.

29. Elías Calles, "Informe rendido por el C. General Plutarco Elías Calles," 245; Sánchez, *Mexico*, 67.

30. "La obra cultural de la estación XFX de la Secretaría Educación Pública," *El Universal*, December 12, 1929, sec. 4, 9.

31. "Aparatos de radio para las escuelas primarias," *El Informador* (Jalapa), September 19, 1929, AHSEP, caja 9474, exp. 25; José Suárez G. to María Luisa Ross, September 24, 1929, AHSEP, caja 9474, exp. 25.

32. José Suárez G. to María Luis Ross, October 22, 1929, AHSEP, caja 9474, exp. 25.

33. "$28,000,000 más para la educación," *La Prensa*, December 8, 1928, CEHM, fondo CCCXXI. carp 7. fold. 566. Over 6,100 rural schools existed in 1929; Booth, *Mexico's School-Made Society*, 114.

34. In addition to the Archivo Histórico de la SEP and U.S. and Mexican newspapers, see Secretaría de Educación Pública, *Las misiones culturales en 1927*; Chávez Ortiz, "La radio educativa en el México revolucionario"; and Sánchez Ruíz, *Orígenes de la radiodifusión en México*, 22.

35. U.S. Department of Commerce, quoted in Sánchez Ruíz, *Orígenes de la radiodifusión*, 22.

36. R. Gandanedo to the Jefe del Departamento, June 17, 1929, AHSEP, caja 9474, exp. 25.

37. "Lista de aparatos receptores adquiridos . . . a cambio de anuncios por la estación X.F.X.," December 1928, AHSEP, caja 9476, exp. 45; Fco. Javier Stávoli, March 7, 1929, AHSEP, caja 9474, exp. 25.

38. Luis F. Rodríguez, "Informe de la visita hecha a las escuelas rurales dotadas de aparato de radio," March 18, 1933, AHSEP, caja 9482, exp. 78; Luis F. Rodríguez, "Informe de la visita hecha a las escuelas rurales dotadas de aparato de radio," March 25, 1933, AHSEP, caja 9482, exp. 78; Luis F. Rodríguez, "Informe de la visita hecha a las escuelas rurales dotadas de aparato de radio, del Edo. de Hidalgo," April 22, 1933, AHSEP, caja 9482, exp. 78; Luis F. Rodríguez, "Informe de la visita de inspección a la Zona de Tlalnepantla, Méx.," April 24, 1933, AHSEP, caja 9482, exp. 78; Luis F. Rodríguez, "Informe de la visita de inspección a la Zona de Texcoco," May 13, 1933, AHSEP, caja 9482, exp. 78.

39. C. E. Sansalvador to Ma. Luisa Ross, Acatlán, Puebla to Mexico City, April 13, 1929, AHSEP, caja 9474, exp. 25.

40. Luis F. Rodríguez, "Informe de la visita hecha a las escuelas rurales dotadas de aparato de radio, del Edo. de Puebla," March 18, 1933, AHSEP, caja 9482, exp. 78; Luis F. Rodríguez, "Informe de la visita hecha a las escuelas rurales dotadas de aparato de radio, del Edo. de Tlaxcala," March 25, 1933, AHSEP, caja 9482, exp. 78.

41. María Luisa Ross, "Informe de las labores desarrolladas por la Obra de Extensión Educativa por Radio de la Secretaría de Educación Pública, durante el mes de septiembre 1929," September 1929, AHSEP, caja 9474, exp. 25.

42. González de Bustamante, *"Muy buenas noches,"* 196–97.

43. González de Bustamante, *"Muy buenas noches,"* 196–97.

44. "Radio Stations in Mexico," *New York Times*, March 11, 1928, 14.

45. Gálvez Cancino "Voz, jinete del aire," 17.

46. "Obregon Slayer on Trial," *Los Angeles Times*, November 3, 1928, 1; "Toral Case Gag Put on Radio and Press, *Washington Post*, November 6, 1928, 5; "Toral and Nun Go on Trial for Murder of General Obregon," *New York Times*, November 3, 1928, 1.

47. F. Javier Stavoli and Alejandro Michel, "Proyecto para la adquisición de una estación de gran potencia para el gobierno de México," Mexico City, April 29, 1930, AHSEP, caja 9475, exp. 29.

48. Velázquez Estrada, "La radiodifusión mexicana," 296–97; Buchenau, *Last Caudillo*, 165.

49. Dulles, *Yesterday in Mexico*, 276.

50. "Propaganda obrera por medio de estaciones radiofónicas," *El Universal*, November 16, 1923, 1; Manuel Azamar to Gral. P. E. Calles, April 12, 1924, APECFT, Archivo Plutarco Elías Calles, exp. 215, inv. 456, leg. 1.

51. Velázquez Estrada, "La radiodifusión mexicana durante los gobiernos de Alvaro Obregón y Plutarco Elías Calles," 156.

52. U.S. Department of Commerce, quoted in Sánchez Ruíz, *Orígenes de la radiodifusión*, 22.

53. Ing. Miguel Fonesca to the Jefe del Departamento, February 4, 1930, AHSEP, caja 9475, exp. 29.

54. "El conflicto de 'El Aguila' tiene seria resonancia en Tampico," *El Universal*, January 1, 1926, 10.

55. "Petroleum Tax Up Slightly in Mexico," *New York Times*, January 29, 1930, 41; "Mexico Oil Sky Clearing," *Los Angeles Times*, February 27, 1928, 16.

56. Meyer, *México y los Estados Unidos*, 270–75; Santiago, *Ecology of Oil*, 282–83.

57. Ing. Miguel Fonesca to the Jefe del Departamento, February 4, 1930, AHSEP, caja 9475, exp. 29; Manuel Azamar to Gral. P. E. Calles, April 12, 1924, APECFT, Archivo Plutarco Elías Calles, exp. 215, inv. 456, leg. 1; José Suárez G. to María Luisa Ross, September 24, 1929, AHSEP, caja 9474, exp. 25; José Suárez G. to María Luisa Ross, October 22, 1929, AHSEP, caja 9474, exp. 25; "Asunto: Lista del material recibido de la Comisión Nacional de Caminos,"1929, AHSEP, caja 9475, exp. 25.

58. "Radioconciertos," *Excélsior*, October 26, 1928, sec. 2, 3.

59. Medina Ávila and Vargas Arana, *Nuestra es la voz*, 163.

60. Gómez Chacón, "Carillo Puerto y la radio en Yucatán," 163–201; Medina Ávila and Vargas Arana, *Nuestra es la voz*, 144; "Mensaje del Gral. Lázaro Cárdenas," *El Universal*, July 1, 1934, 1, 3.

61. "Seguirá el descuento a los empleados para el P.N.R.," *La Prensa*, October 9, 1930, CEHM, fondo CCXII, carp 24, fold. 231.

62. Mejía Prieto, *Historia de la radio y la t.v.*, 55; "Resultó un éxito el radio-concierto," *El Nacional*, June 29, 1929, 1, 6.

63. B. R. Armstrong, "Propagandists in Mexico Busy," *Los Angeles Times*, October 9, 1929, 2.

64. "Reds Seize Radio Station," *New York Times*, November 9, 1931, 37.

65. Hayes, *Radio Nation*, 39.

66. Albarrán, *Seen and Heard*, 141.

67. Alejandro Michel, "Recopilación detallada de los servicios que presta la obra de Extensión Educativa por Radio de la SEP," Mexico City, October 29, 1930, AHSEP, caja 9478, exp. 4.

68. Departamento de Prensa y Publicidad, "Partido Nacional Revolucionario," 1933, AHSEP, caja 9485, exp. 63; Dr. J. M. Puig Casauranc to Senador Don Silvestre Guerrero, January 19, 1931, AHSEP, caja 9480, exp. 35; María Luisa Ross to Silvestre Guerrero, January 21, 1931, AHSEP, caja 9480, exp. 35.

69. Mejía Prieto, *Historia de la radio y t.v.*, 57.

70. "Resultó un éxito el radio-concierto," *El Nacional*, June 29, 1929, 1, 6.

71. "Ortiz Rubio Speaks on Radio to Mexico," *New York Times*, February 27, 1930, 6.

72. "Un mensaje de buenos deseos para el pueblo de la nación," *Excélsior*, January 1, 1931, 1, 3.

73. Manuel Jasso, quoted in Mejía Prieto, *Historia de la radio y t.v.*, 56.

74. "Seguirá el descuento a los empleados para el P.N.R.," *La Prensa*, October 22, 1930, CEHM, fondo CCXII, carp. 24, fold. 231.

75. María Luisa Ross, "Informe de las labores . . . durante el periodo comprendido del 1 de agosto de 1930 al 21 de Julio de 1931," July 31, 1931, AHSEP, caja 9478, exp. 4.

76. "New Radio Voice," *Los Angeles Times*, January 11, 1934, A4.

77. "Transmisión del informe a toda la República," *Excélsior*, September 1, 1934, 4.

78. Norman Baker to Plutarco Elías Calles, April 23, 1934, APECFT, Archivo Plutarco Elías Calles, exp. 7, inv. 458, leg. 1.

79. Hayes, *Radio Nation*, 86.

80. "Seguirá el descuento a los empleados para el P.N.R.," *La Prensa*, October 2, 1930, CEHM, fondo CCXII, carp. 24, fold. 231.

81. Lázaro Cárdenas, *El Sr. Gral de División Lázaro Cárdenas: Candidato a la Presidencia de la Republica por el P.N.R., hace profesión de fe cooperativista* (Mexico City: Liga Nacional de Acción Cooperativa, 1934): 1–4, APECFT, Fondo Presidentes, exp. 5, inv. 803, leg. 1.

82. "Mexico City Adopts Hush Regulation," *Los Angeles Times*, December 20, 1931, 1.

83. "Mexico City Acts to Cut Noise, Aiming especially at Radio," *New York Times*, March 23, 1931, 8.

84. Andrés Figueroa, in *Memoria por la Secretaría de Comunicaciones y Obras Públicas, 1935–1936*, 90–96.

85. Francisco J. Múgica, in *Memoria por la Secretaría de Comunicaciones y Obras Públicas, 1937–1938*, 13, 25.

86. Múgica, in *Memoria por la Secretaría de Comunicaciones y Obras Públicas, 1936–1937*, 23–27; Múgica, in *Memoria por la Secretaría de Comunicaciones y Obras Públicas, 1937–1938*, 28.

87. Melquiades Angulo, in *Memoria por la Secretaría de Comunicaciones y Obras Públicas, 1939–1940*, 117.

88. Instituto Politecnico Nacional, Generación 70–74, "Don José R. de la Herran," 9–13.

89. Méndez Docurro, *Gral. Guillermo Garza Ramos y Trillo*, 77–80.

90. Medina Ávila and Vargas Arana, *Nuestra es la voz*, 132; Herbert Cerwin to John C. Royal, c. 1943, WHS, NBC Collection, Royal Papers, box 111, fold. 48.

CONCLUSION

1. Lázaro Cárdenas, "Expropiación de la industria petrolera: El Sr. Presidente dirige transcendental mensaje al pueblo de la república," *El Nacional*, March 19, 1938, 1; Niblo, *War, Diplomacy, and Development*, 36; Hayes, *Radio Nation*, 83–85; Santiago, *Ecology of Oil*, 338.

2. For more on Cedillo, see Ankerson, *Agrarian Warlord*; Navarro, *Political Intelligence*, 262–65.

3. Niblo, *War, Diplomacy, and Development*, 45–47.

4. "Como se comunicaba el General Cedillo," *El Universal*, May 21, 1938, 1, 10; Medina Ávila and Vargas Arana, *Nuestra es la voz*, 176–77.

5. Wu, *Master Switch*, 5–6.

6. Linguist and media critic Noam Chomsky has addressed this subject of media consolidation in the United States for years. See Herman and Chomsky, *Manufacturing Consent*; Chomsky, *Necessary Illusions: Thought Control in Democratic Societies*; Chomsky and Barsamian, *Propaganda and the Public Mind: Conversations with Noam Chomsky*.

7. Wu, *Master Switch*, 39. Exceptions to this finding exist in the rise of local community stations; see, e.g., Rodríguez, *Citizens' Media against Armed Conflict*.

8. See Neulander, *Programming National Identity*; Claxton, *From Parsifal to Perón*; McCann, *Hello, Hello Brazil*. Britain and Germany followed more state-controlled approaches to broadcasting, whereas the United States pushed a private-enterprise system.

9. McCann, *Hello, Hello Brazil*, 23.

10. Lawson, *Building the Fourth Estate*, 18; Bustamante, *"Muy buenas noches,"* 5–15.

11. Lawson, *Building the Fourth Estate*, 30. A *priista* is a supporter of the PRI.

12. Lawson, *Building the Fourth Estate*, 30.

13. Jo Tuckman, "WikiLeaks reveals US Concerns over Televisa-Peña Nieto Links in 2009," *The Guardian*, June 11, 2012, http://www.guardian.co.uk/world/2012/jun/11/wikileaks-us-concerns-televisa-pena-nieto, accessed December 28, 2012; Jo Tuckman, "Mexican Media Scandal," *The Guardian*, June 26, 2012, http://www.guardian.co.uk/world/2012/jun/26/mexican-media-scandal-televisa-pri-nieto, accessed 28 December 2012; Jenaro Villamil, "Proyecto Jorge: El plan Televisa-Peña Nieto para alcanzar la presidencia," *Proceso*, September 8, 2012, http://www.proceso.com.mx/?p=319353, accessed December 28, 2012. There were numerous other reasons for Peña Nieto's victory, including security concerns,

growing animosity toward the rule of the rightist Partido Acción Nacional (PAN), and fears that the leftist candidate Andrés Manuel López Obrador would be destabilizing if in presidential office.

14. Niblo, *War, Diplomacy, and Development*, 35–57.
15. "Comentemos el 'encadenamiento,'" *El Nacional*, December 13, 1939, 7.

BIBLIOGRAPHY

ARCHIVAL SOURCES

AGEY   Archivo General de la Estado de Yucatán
AGN   Archivo General de la Nación de México
AHDF   Archivo Histórico del Distrito Federal
AHI   Universidad Iberoamericana, Acervos Históricos
AHSDN   Archivo Histórico de la Secretaría de la Defensa Nacional
AHSEP   Archivo Histórico de la Secretaría de Educación
APECFT   Fideicomiso Archivos Plutarco Elías Calles y Fernando Torreblanca
APS   American Periodicals Series
ASRE   Archivo Histórico Genaro Estrada de la Secretaría de Relaciones Exteriores
BMLT   Biblioteca Miguel Lerdo de Tejada, Archivo Económicos
BTTSCT   Biblioteca Telecomunicaciones y Telegrafía de la Secretaría de Comunicaciones y Transportes
CEHM   Centro de Estudios de Historia de México, CARSO
NARA   U.S. National Archives and Record Administration
NLBL   Nettie Lee Benson Library, University of Texas at Austin
WHS   Wisconsin Historical Society

PUBLISHED SOURCES

Aceves González, Francisco de Jesús, Pablo Arredondo, and Carlós Luna, comps. *Radiodifusión regional en México: Historias, programas, audiencias.* Guadalajara: Universidad de Guadalajara, 1989.

Adas, Michael. *Machines as the Measure of Men: Science, Technology, and Ideology of Western Dominance.* Ithaca NY: Cornell University Press, 1989.

Agostoni, Claudia. *Monuments of Progress: Modernization and Public Health in Mexico City, 1876–1910.* Calgary: University of Calgary Press, 2003.

Ahvenainen, Jorma. *The European Cable Companies in South America before the First World War.* Helsinki: Finnish Academy of Science and Letters, 2004.

——— . *The History of the Caribbean Telegraphs before the First World War*. Helsinki: Finnish Academy of Science and Letters, 1996.

Aitken, Hugh G. J. *Syntony and Spark: The Origin of Radio*. 2nd ed. Princeton NJ: Princeton University Press, 1985.

Albarrán, Elena Jackson. "Children of the Revolution: Constructing the Mexican Citizen, 1920–1940." PhD dissertation, University of Arizona, 2008.

——— . *Seen and Heard in Revolutionary Mexico: Children and Cultural Nationalism, 1921–1940*. Lincoln: University of Nebraska Press, 2014.

Albert, Pierre, and André-Jean Tudesq. *Historia de la radio y la televisión*. Mexico City: Fondo de Cultura Económico, 1982.

Alisky, Marvin. "Early Mexican Broadcasting." *Hispanic American Historical Review* 34, no. 4 (November 1954): 513–26.

——— . "Educational Aspects of Broadcasting in Mexico." PhD dissertation, University of Texas, 1953.

——— . "Mexico's Rural Radio." *Quarterly of Film Radio and Television* 8, no. 4 (Summer 1954): 405–17.

Alva de la Selva, Alma Rosa. *Radio e ideología*. 2nd ed. Mexico City: Ediciones El Caballito, 1982.

Alvarado Mendoza, Arturo. *El Portesgilismo en Tamaulipas: Estudio sobre la constitución de la autoridad pública en el Mexico posrevolucionario*. Mexico City: Colegio de México, 1992.

Anda Gutiérrez, Cuauhtémoc. *Importancia de la radiodifusión en México*. Mexico City: Luis Cabrera, 2004.

Anderson, Benedict. *Imagined Communities: Reflections on the Origin and Spread of Nationalism*. London: Verso, 1983.

Anduaga, Aitor. *Wireless and Empire: Geopolitics, Radio Industry and Ionosphere in the British Empire, 1918–1939*. Oxford: Oxford University Press, 2009.

Ankerson, Dudley. *Agrarian Warlord: Saturnino Cedillo and the Mexican Revolution in San Luis Potosí*. DeKalb: Northern Illinois University Press, 1984.

Antonio de Noriega, Luis, and Frances Leach. *Broadcasting in Mexico*. London: Routledge & Kegan Paul, 1979.

Archer, Gleason. *The History of the Radio to 1926*. New York: American Historical Society, 1938.

Asturias, Miguel Ángel. *El señor presidente*. Madrid: Cátedra, 1997.

Baker, W. J. *History of the Marconi Company*. New York: St. Martin's Press, 1970.

Balk, Alfred. *The Rise of Radio: From Marconi through the Golden Age*. Jefferson NC: McFarland, 2006.

Banti, Angelo. *Il telefono senza fili sistema Marconi*. Roma: Gli Editori Dell'*Elettricistta*, 1897.

Bantjes, Adrian A. *As if Jesus Walked on Earth: Cardenismo, Sonora, and the Mexican Revolution*. Lanham MD: SR Books, 1998.

Barbero, Raúl E. *De la galena al satélite: Crónica de 70 años de radio el Uruguay, 1922–1992.* Uruguay: Ediciones de la Pluma, 1995.

Barbour, Philip. "Commercial and Cultural Broadcasting in Mexico." In "Mexico Today," special issue, *Annals of the American Academy of Political and Social Science* 208 (March 1940): 94–102.

Barragán Rodríguez, Juan. *Historia del ejército y la revolución constitucionalista.* Vol. 2. Mexico City: Talleres de la Editorial Stylo, 1946.

Barty-King, Hugh. *Girdle Round the Earth.* London: Heinman, 1979.

Batson, Lawrence D. *Radio Markets of the World, 1926, 1928–1929, 1930, 1932.* Washington DC: Government Printing Office, 1926, 1929, 1930, 1932.

Beezley, William. *Judas at the Jockey Club.* 2nd ed. Lincoln: University of Nebraska Press, 2001.

Benbow, Mark. *Leading Them to the Promised Land: Woodrow Wilson, Covenant Theology, and the Mexican Revolution, 1913–1915.* Kent OH: Kent University Press, 2010.

Benjamin, Thomas. *La Revolución: Mexico's Great Revolution as Memory, Myth, and History.* Austin: University of Texas Press, 2000.

Betancourt, Enrique C. *Apuntes para la historia: Radio, televisión y farándula de la Cuba de ayer.* San Juan PR: Ramallo Brothers, 1966.

Booth, George C. *Mexico's School-Made Society.* Stanford CA: Stanford University Press, 1941.

Bravo Izquierdo, Donato. *Un soldado del pueblo.* Mexico City: n.p., 1964.

The Brinkley Hospital for the Treatment of Rectal and Colonic Disorders, Varicose Veins and Ulcers, Hernia or Rupture. Little Rock AR: Brinkley Hospitals, 1939.

Britton, John A. "'The Confusion Provoked by Instantaneous Discussion': The New International Communications Network and the Chilean Crisis of 1891–1892 in the United States." *Technology and Culture,* no. 48 (October 2007): 729–57.

Britton, John A., and Jorma Ahvenainen. "Showdown in South America: James Scrymser, John Pender, and United States–British Cable Competition." *Business History Review* 78 (Spring 2004): 1–27.

Brock, Pope. *Charlatan: America's Most Dangerous Huckster, the Man Who Pursued Him, and the Age of Flimflam.* New York: Crown, 2008.

Bronfman, Alejandra, and Andrew Grant Wood, eds. *Media, Sound, and Culture in Latin America and the Caribbean.* Pittsburgh PA: University of Pittsburgh Press, 2012.

Brown, F. J. *The Cable and Wireless Communications of the World.* London: Sir Isaac Pittman & Sons, 1927.

Brunk, Samuel. *¡Emiliano Zapata! Revolution and Betrayal in Mexico.* Albuquerque: University of New Mexico Press, 1995.

Buchenau, Jürgen. *In the Shadow of the Giant: The Making of Mexico's Central American Policy, 1876–1930.* Tuscaloosa: University of Alabama Press, 1996.

———. *The Last Caudillo: Alvaro Obregón and the Mexican Revolution.* Chichester, UK: Wiley-Blackwell, 2011.

———. *Plutarco Elías Calles and the Mexican Revolution.* Lanham MD: Rowman & Littlefield, 2006.

Bulmer-Thomas, Victor. *The Economic History of Latin America since Independence.* Cambridge: Cambridge University Press, 1994.

Bunker, Steve B. *Creating Mexican Consumer Culture in the Age of Porfirio Díaz.* Albuquerque: University of New Mexico Press, 2012.

Burns, Russell. *Communications: An International History of the Formative Years.* London: Institution of Electrical Engineers, 2004.

Cámara Nacional de la Industria de la Radio y Televisión. *Cápsulas culturales: Espacios abiertos en radio y televisión a las raíces de México.* Mexico City: Cámara Nacional de la Industria de la Radio y Televisión, 1985.

Campbell, Timothy C. *Wireless Writing in the Age of Marconi.* Minneapolis: University of Minnesota Press, 2006.

Cantril, Hadley, and Gordon W. Allport. *The Psychology of Radio.* New York: Harper & Brothers, 1971.

Cárdenas de la Pena, Enrique. *El telégrafo.* Mexico City: Secretaría de Comunicaciones y Transportes, 1987.

———. *Semblanza marítima del México independiente y revolucionario.* Vol. 2. Mexico City: Secretaría de Marina, 1970.

Carrillo Ana María. "¿Estado de pete o estado de sitio?: Sinaloa y Baja California, 1902–1903." *Historia Mexicana* 54, no. 4 (April–June 2005): 1049–1103.

Carrillo Olano, Alejandra. "Radio Altiplano del estado Tlaxcala: Entre el modelo comercial y la radio pública." Tesis profesional, Universidad de las Américas Puebla, 2007. http://catarina.udlap.mx/u_dl_a/tales/documentos/lco/carrillo_o_a/index.html. Accessed February 16, 2011.

Carson, Gerald. *The Roguish World of Doctor Brinkley.* New York: Holt, Rinehart and Winston, 1960.

Castro, J. Justin. "Radiotelegraphy to Broadcasting: Wireless Communications in Porfirian and Revolutionary Mexico, 1899–1924." *Mexican Studies / Estudios Mexicanos* 29, no. 2 (Summer 2013): 335–65.

———. "Sounding the Mexican Nation: Intellectuals, State Building, and the Culture of Early Radio Broadcasting." *Latin Americanist* 58, no. 3 (September 2014): 3–30.

Castro Martínez, Pedro. *Adolfo de la Huerta y la Revolución Mexicana.* Mexico City: INEHRM/UAM, 1990.

———. *Álvaro Obregón: Fuego y cenizas de la Revolución Mexicana.* Mexico City: Ediciones Era, 2009.

Chávez Ortiz, Ivonne Grethel. "La radio educativa en el México revolucionario." Tesis de licenciatura, Universidad Autónoma Metropolitana, 2001.

Chomsky, Noam. *Necessary Illusions: Thought Control in Democratic Societies.* Boston: South End Press, 1989.

Chomsky, Noam, and David Barsamian. *Propaganda and the Public Mind: Conversations with Noam Chomsky.* Cambridge MA: South End Press, 2001.

Claxton, Robert Howard. *From Parsifal to Perón: Early Radio in Argentina, 1920–1944.* Gainesville: University Press of Florida, 2007.

Clegern, Wayne M. "British Honduras and the Pacification of Yucatan." *Americas* 18, no. 3 (January 1962): 243–54.

Coatsworth, John H. *Growth against Development: The Economic Impact of Railroads in Porfirian Mexico.* DeKalb: Northern Illinois Press, 1981.

Coatsworth, John H., and Alan M. Taylor, eds. *Latin America and the World Economy since 1800.* Cambridge MA: Harvard University Press, 1998.

Coe, Lewis. *Wireless Radio: A Brief History.* Jefferson NC: McFarland, 1996.

Collins, A. Frederick. "The Slaby-Arco Portable Field Equipment for Wireless Telegraphy." *Scientific American* 85, no. 26 (December 1901): 425–26.

Comisión Inter-Americana de comunicaciones eléctricas: Convención, resoluciones y actas. Mexico City: Gobierno de los Estados Unidos Mexicanos, 1926.

Connolly, Priscilla. *El contratista de don Porfirio: Obras públicas, deuda y desarrollo desigual.* Mexico City: Fondo de Cultura Económica, 1997.

"Continuaciones de 'el estudio.'" 1, no. 1. *Anales de Instituto Médico Nacional.* Mexico City: Oficina Tipográfica de la Secretaría de Fomento, 1894.

Contreras, Mario, and Jesús Tamayo, eds. *México en siglo XX, 1900–1913: Textos y documentos.* 2 vols. Mexico City: Universidad Nacional Autónoma de México, 1983.

Coronado, Jorge. *The Andes Imagined: Indigenismo, Society, and Modernity.* Pittsburgh: University of Pittsburgh Press, 2009.

Coronado Ponce, Alán René. "La radiodifusión familiar en México y su inserción en la dinámica de concentración de medios: Un estudio de caso de Guadalajara." Tesis de maestría, Universidad de Guadalajara, 2004.

Craib, Raymond B. *Cartographic Mexico: A History of State Fixations and Fugitive Landscapes.* Durham NC: Duke University Press, 2004.

Crookes, William. "Address of the President before the British Association for the Advancement of Science, Bristol, 1898." *Science,* n.s., 8, no. 201 (November 4, 1898): 561–75.

Cue Canovas, Agustín. *Ricardo Flores Magón: La Baja California y Los Estados Unidos.* Mexico City: Libro Mex, 1957.

Curiel, Fernando. *¡Dispara margot, dispara! Un reportaje justiciero de la radio difusión mexicana.* Mexico City: Premia Editora, 1987.

———. *La telaraña magnética y otros estudios radiofónicos.* Mexico City: Ediciones Coyoacán, 1997.

D'Agostino, Salvo. "Hertz's Researches on Electromagnetic Waves." *Historical Studies in the Physical Sciences* 6 (1975): 261–323.

Daniels, Josephus. *Shirt Sleeve Diplomat*. Chapel Hill: University of North Carolina Press, 1947.

de Andrade Martins, Roberto. "Resistance to the Discovery of Electromagnetism: Ørsted and the Symmetry of the Magnetic Field," 165–85. http://ppp.unipv. it/Collana/Pages/Libri/Saggi/Volta%20and%20the%20History%20of%20 Electricity/V%26H%20Sect3/V%26H%20245-265.pdf. Accessed January 21, 2012.

de Armas Chitty, J. A. *Historia de la radiodifusión en Venezuela*. Caracas: Edición de la Cámara Venezolana de la Industria de la Radiodifusión, 1975.

de Dios Bonilla, Juan. *Apuntes para la historia de la marina nacional*. Mexico City: n.p., 1946.

de Fornaro, Carlo, ed. *Carranza and Mexico*. New York: Mitchell Kennerley, 1915.

Delgadillo García, Octavio. "Radio Broadcasting and Popular Culture: Forming the Nation in Oaxaca, Mexico, 1920–1940." Masters thesis, San Diego State University, 2007.

de los Reyes, Aurelio. *Medio siglo de cine mexicano (1896–1947)*. Mexico City: Editorial Trillas, 1987.

——, ed. *Historia de la vida cotidiana en México*. Vol. 5, *Siglo XX: Campo y ciudad*. Mexico City: Colegio de México / Fondo de Cultura Económica, 2006.

Delpar, Helen. *The Enormous Vogue of Things Mexican: Cultural Relations between the United States and Mexico, 1920–1935*. Tuscaloosa: University of Alabama Press, 1995.

——. "Mexican Culture, 1920–1945." In *The Oxford History of Mexico*, edited by Michael C. Meyer and William H. Beezley, 543–72. Oxford: Oxford University Press, 2000.

Diacon, Todd A. *Stringing Together a Nation: Cândido Mariano de Silva Rondon and the Construction of a Modern Brazil, 1906–1930*. Durham NC: Duke University Press, 2004.

Douglas, Susan J. *Inventing American Broadcasting, 1899–1922*. Baltimore MD: John Hopkins University Press, 1987.

Dowsett, H. M. *Wireless Telephony and Broadcasting*. London: Gresham, 1924.

Dulles, John W. F. *Yesterday in Mexico: A Chronicle of the Revolution, 1919–1936*. Austin: University of Texas, 1961.

Eccles, W. H. *Wireless*. London: Thornton Butterworth, 1933.

Eisenhower, John S. D. *Intervention! The United States and the Mexican Revolution, 1913–1917*. New York: W. W. Norton, 1995.

Elías Calles, Plutarco. *Pensamiento político y social: Antología (1913–1936)*. Mexico City: Fondo de Cultura Económica, 1988.

England, Shawn Louis. "The Curse of Huitzilopochtli: Origins, Process, and Legacy of Mexico's Military Reforms, 1920–1946." PhD dissertation, Arizona State University, 2008.

Esparza, Rafael R. *La aviación*. Mexico City: Secretaría de Comunicaciones y Transportes, 1987.

Fallaw, Ben. "Eulogio Ortiz: The Army and the Antipolitics of Postrevolutionary State Formation." In *Forced Marches: Soldiers and Military Caciques in Modern Mexico*, edited by Ben Fallaw and Terry Rugeley. Tucson: University of Arizona Press, 2012.

Felicitas Arias, Elisa. "The Metrology of Time." *Philosophical Transactions: Mathematical, Physical, and Engineering Sciences* 363, no. 1834 (September 2005): 2290–91.

Fell, Claude. *José Vasconcelos: Los años del águila*. Mexico City: Universidad Nacional Autónoma de México, 1989.

Fernández Christlieb, Fátima. *La radio mexicana: Centro y regiones*. Mexicali: Juan Pablos, 1991.

———. *Los medios de difusión masiva en México*. Mexico City: Juan Pablos, 1982.

Fernández Ramírez, J., and José de la Herrán. "Radio: Nuestra experiencia en radio." *Revista Mexicana de Ingeniería y Arquitectura* 1, nos. 7–10 (September–December 1923): 430–35, 527–34, 599–605, 672–81.

Figueroa Bermúdez, Romeo. *¡Qué onda con la radio!*. Mexico City: Pearson Educación, 1997.

Filine, Benjamin. *Romancing the Folk: Public Memory and American Roots Music*. Chapel Hill: University of North Carolina Press, 2000.

Finn, Bernard, and Daqing Yang, eds. *Communications under the Seas*. Cambridge MA: Massachusetts Institute of Technology Press, 2009.

Flores Magón, Ricardo. *Dreams of Freedom: A Ricardo Flores Magón Reader*, edited by Chaz Bufe and Mitchell Cowen Verter. Edinburgh: AK Press, 2005.

Flower, Sydney B., ed. *The Goat-Gland Transplantation: As Originated and Successfully Performed by J. R. Brinkley, M.D., of Milford, Kansas U.S.A., in Over 600 Operations upon Men and Women*. Chicago: New Thought Book Department, 1921.

Fowler, Gene, and Bill Crawford. *Border Radio: Quacks, Yodelers, Pitchmen, Psychics, and Other Amazing Broadcasters of the American Airwaves*. Austin: University of Texas Press, 2002.

Fox, Elizabeth. *Latin American Broadcasting: From Tango to Telenovela*. Lutin UK: Lutin University Press, 1997.

Frank, Patrick. *Posada's Broadsheets: Mexican Popular Imagery, 1890–1910*. Albuquerque: University of New Mexico Press, 1998.

Fuentes, Gloria. *La radiodifusión*. Mexico City: Secretaría de Comunicaciones y Transportes, 1988.

Fuentes Díaz, Vicente. *Historia de la Revolución en el Estado de Guerrero*. Mexico City: Instituto Nacional de Estudios Históricos de la Revolución Mexicana, 1983.

Gallo, Rubén. *Mexican Modernity: The Avant-Garde and the Technological Revolution*. Cambridge: Massachusetts Institute of Technology Press, 2005.

Gálvez Cancino, Felipe. "Los felices del alba." Tesis de licenciatura, Universidad Nacional Autónoma de México, 1975.

———."Voz, jinete del aire." *México en el tiempo. Revista de Historia y Conservación* 3, no. 23 (March-April 1998): 15–17.

Gargurevich, Juan. *La Peruvian Broadcasting Co.: Historia de la radio.* Lima: La Voz Ediciones, 1995.

Garner, Paul. *Porfirio Díaz.* Edinburgh: Longman, 2001.

Gastélum, Bernardo J. *Palabras del Dr. Bernardo J. Gastélum en la inauguración de la estación de radio de la secretaria de educación pública C. Y. E., instalada en esa dependencia del ejecutivo por acuerdo del C. Secretario de Educación Dr. Bernardo J. Gastélum, siendo Presidente de la Republica el C. Gral. Alvaro Obregón México, 30 de noviembre de 1924.* Mexico City: Editorial "cultura," 1924.

"General Postal Union; October 9, 1874," Avalon Project: Documents in Law, History and Diplomacy, Yale Law School, Lillian Goldman Law Library. http://avalon.law.yale.edu/19th_century/usmu010.asp. Accessed November 22, 2010.

Gilly, Adolfo. *The Mexican Revolution.* Translated by Patrick Camiller. New York: New Press, 2005.

Glazebrook, Richard. "The Origins of Wireless." *Scientific Monthly* 20, no. 3 (March 1925): 291–96.

Gómez Chacón, Gaspar. "Carillo Puerto y la radio en Yucatán." In *La Revolución en Yucatán: Nuevos ensayos,* 163–201. Mérida: Secretaría de Educación, CESPA Editorial, 2012.

Gómez Vargas, Héctor. *Memorias suspendidas: Orígenes de la radio en León.* León: Consejo para la Cultura de León y Universidad Iberoamericana, 1998.

Gonzales, Michael J. "Imagining Mexico in 1921: Visions of the Revolutionary State and Society in the Centennial Celebration in Mexico City," 25, no. 2 (Summer 2009): 247–70.

———. *The Mexican Revolution, 1910–1940.* Albuquerque: University of New Mexico Press, 2002.

González, Manuel W. *Contra Villa.* Mexico City: Editorial Botas, 1935.

González Cruz, Edith. *La compañía El Boleo: Su impacto e la municipalidad de Mulegé, 1885–1918.* La Paz: Universidad Autónoma de Baja California Sur, 2000.

González de Bustamante, Celeste. *"Muy buenas noches": Mexico, Television, and the Cold War.* Lincoln: University of Nebraska Press, 2013.

Granados, Pával. *XEW: 70 años en el aire.* Mexico City: Editorial Clio, 2000.

Grunstein Dicter, Arturo. "¿Nacionalista porfiriano o 'científico extranjerista'? Limantour y la consolidación ferroviaria en la crisis del antiguo régimen y el estallido de la revolución." In Leyva, Gustavo, et al., eds. *Independencia y revolución: Pasado, presente y futuro.* Mexico City: Universidad Autónoma Metropolitana / Fondo de Cultura Económica, 2010.

Guzmán Cantú, Tomás. "Telegrafía sin hilos." In *Historia de las telecomunicacio-nes*, edited by Manuel Rosales Vargas and Virginia Licona Peña. Mexico City: Telecomm / Telégrafos, 1999.

Haber, Stephen. *Industry and Underdevelopment: The Industrialization of Mexico, 1900–1940*. Stanford CA: Stanford University Press, 1989.

Haber, Stephen, Armando Razo, and Noel Maurer. *The Politics of Property Rights: Political Instability, Credible Commitments, and Economic Growth, 1876–1929* Cambridge MA: Cambridge University Press, 2003.

Hagedom, Dan. *Conquistadors of the Sky: A History of Aviation in Latin America.* Gainesville: University Press of Florida, 2008.

Halperín Donghi, Tulio. *The Contemporary History of Latin America*. Edited and translated by John Charles Chasteen. Durham NC: Duke University Press, 1993.

Hamnett, Brian. *A Concise History of México*. 2nd ed. Cambridge: Cambridge University Press, 2006.

Harbord, James G. "America's Position in Radio Communications." *Foreign Affairs* 4, no. 3 (April 1926): 465–74.

Hart, John M. *Empire and Revolution: The Americans in Mexico since the Civil War.* Berkeley: University of California Press, 2002.

Hart, Paul. *Bitter Fruit: The Social Transformation of Morelos, Mexico, and the Origins of the Zapatista Revolution, 1840–1910*. Albuquerque: University of New Mexico Press, 2005.

Hayes, Joy. "National Imaginings on the Air: Radio in Mexico, 1920–1950." In *The Eagle and the Virgin: Nation and Culture in Mexico, 1920–1940*, edited by Mary Kay Vaughn and Stephen E. Lewis, 243–58. Durham NC: Duke University Press, 2006.

———. *Radio Nation: Communication, Popular Culture, and Nationalism in Mexico, 1920–1950*. Tucson: University of Arizona Press, 2000.

Headrick, Daniel R. *The Invisible Weapon: Telecommunications and International Politics, 1851–1945*. New York: Oxford University Press, 1991.

———. *The Tentacles of Progress: Technology Transfer in the Age of Imperialism, 1850–1914*. New York: Oxford University Press, 1988.

———. *The Tools of Empire: Technology and European Imperialism in the Nineteenth Century*. New York: Oxford University Press, 1981.

Hearings on the International Radio Telegraph Convention, Before the Committee of Foreign Relations, U.S. Senate, 17th Congress, 1st Session. Washington DC: Government Printing Press, 1928.

Hearings on Remote Control Border Stations—H.R. 7800, Before the Committee on Merchant Marine, Radio, and Fisheries, House of Representatives, 73rd Congress, 2nd Session (February 15, 1934) (statement of C. B. Joliffe, Chief Engineer, Federal Radio Commission).

Henderson, Peter. *In the Absence of Don Porfirio: Francisco León de la Barra and the Mexican Revolution.* Wilmington DE: SR Books, 2000.

Herman, Edward S., and Noam Chomsky. *Manufacturing Consent: The Political Economy of the Mass Media.* New York: Pantheon Books, 1988.

Hills, Jill. *The Struggle for Control of Global Communications: The Formative Century.* Urbana: University of Illinois Press, 2002.

Hobsbawm, Eric. *Nations and Nationalism since 1780: Programme, Myth, Reality.* Cambridge: Cambridge University Press, 1991.

Hochschild, Adam. *King Leopold's Ghost: A Story of Greed, Terror, and Heroism in Colonial Africa.* New York: First Mariner Books, 1999.

Hong, Sungook. "Marconi and the Maxwellians: The Origins of Wireless Telegraphy Revisited." *Technology and Culture* 35, no. 4 (October 1994): 717–49.

———. *Wireless: From Marconi's Black-Box to the Audion.* Cambridge MA: Massachusetts Institute of Technology Press, 2001.

Howard, Philip N., and Muzammil M. Hussain. *Democracy's Fourth Wave?: Digital Media and the Arab Spring.* New York: Oxford University Press, 2013.

Howeth, Linwood S. *History of Communications-Electronics in the United States Navy.* Lansing: University of Michigan Library, 1963.

Hugill, Peter. *Global Communications since 1844: Geopolitics and Technology.* Baltimore MD: Johns Hopkins University, 1999.

Hurtado, Albert L. "Empires, Frontiers, Filibusters, and Pioneers: The Transnational World of John Sutter." *Pacific Historical Review* 77, no. 1 (February 2008): 19–47.

Huuderman, Anton A. *The Worldwide History of Telecommunications.* Hoboken NJ: Wiley-Interscience, 2003.

Informes de las dependencias de la Secretaría de Comunicaciones y Obras Públicas del 11 de abril de 1919 al 31 de mayo de 1920. Mexico City: Dirección de Talleres Gráficos, 1921.

Innis, Harold. *Empire and Communications.* Lanham MD: Rowman & Littlefield, 2007.

Instituto Politecnico Nacional, Generación 70–74. "Don José R. de la Herran, Pionero de la Radiodifusión mexicana." Mexico City: Asociación Mexicana de Periodismo Cientifico, 1983.

Inter-American Committee on Electrical Communications: City of Mexico, May 27–July 22, 1924. Mexico City: Secretaría de Relaciones Exteriores, 1926.

International Radiotelegraph Conference of Washington, 1927. Washington DC: Government Printing Office, 1928.

International Radio Telegraph Convention of Berlin, 1906. Washington DC: Government Printing Press, 1912.

International Radiotelegraph Convention of Washington, 1927. London: His Majesty's Stationary Office, 1928.

Javier Mora, Francisco, ed. *El ruido de las nueces: List Arzubide y el estridentismo mexicano*. Salamanca: Universidad de Alicante, 1999.

Joseph, Gilbert M. *Revolution from Without: Yucatán, Mexico, and the United States, 1880–1924*. Durham NC: Duke University Press, 1988.

Joseph, Gilbert M., and Daniel Nugent, eds. *Everyday Forms of State Formation: Revolution and the Negotiation of Rules in Modern Mexico*. Durham NC: Duke University Press, 1994.

Juhnke, Eric S. *Quacks and Crusaders: The Fabulous Careers of John Brinkley, Norman Baker, and Harry Hoxsey*. Lawrence: University of Kansas Press, 2002.

Kahn, Ed. "The Carter Family on Border Radio." *American Music* 14, no. 2 (Summer 1996): 205–17.

Karush, Mathew B. *Culture of Class: Radio and Cinema in the Making of a Divided Argentina, 1920–1946*. Durham NC: Duke University Press, 2012.

Katz, Friedrich. *The Life and Times of Pancho Villa*. Stanford CA: Stanford University Press, 1998.

———. *The Secret War in Mexico: Europe, the United States, and the Mexican Revolution*. Chicago: University of Chicago Press, 1984.

Keegan, John. *Intelligence in War: Knowledge of the Enemy from Napoleon to Al-Qaeda*. New York: Alfred K. Knopf, 2003.

Kiddle, Amelia M., and María L. O. Muñoz, eds. *Populism in Twentieth Century Mexico: The Presidents of Lázaro Cárdenas and Luis Echeverría*. Tucson: University of Arizona Press, 2010.

King, Clyde L., ed. "Legislative Notes and Reviews." *American Political Science Review* 24, no. 3 (August 1930): 659–65.

King Cobos, Josefina. *Memorias de radio UNAM, 1937–2007*. Mexico City: Universidad Nacional Autónoma de México, 2007.

Knight, Alan. *The Mexican Revolution*. 2 vols. Lincoln: University of Nebraska Press, 1986.

———. "Popular Culture and the Revolutionary State in Mexico, 1910–1940." *Hispanic American Historical Review* 74, no. 3 (August 1994): 393–444.

———. "Populism and Neo-populism in Latin America, Especially Mexico." *Journal of Latin American Studies* 30, no. 2 (May 1998): 223–48.

Krauze, Enrique. *Mexico, Biography of Power: A History of Modern Mexico, 1810–1996*. Translated by Hank Heifetz. New York: HarperPerennial, 1998.

Krysko, Michael A. *American Radio in China: International Encounters with Technology and Communications, 1919–1941*. New York: Palgrave Macmillan, 2011.

La educación pública en México: A través de los mensajes presidenciales desde la consumación de la independencia hasta nuestros días. Mexico City: Publicaciones de la Secretaría de Educación, 1926.

Lawson, Chappell H. *Building the Fourth Estate: Democratization and the Rise of a Free Press in Mexico*. Berkeley: University of California Press, 2002.

Leal, Luis. "Torres Bodet y los 'Contemporáneos.'" *Hispania* 40, no. 3 (September 1957): 290–96.

Lear, John. *Workers, Neighbors, and Citizens: The Revolution in Mexico City.* Lincoln: University of Nebraska Press, 2001.

Lee, R. Alton. *The Bizarre Careers of John R. Brinkley.* Louisville: University of Kentucky Press, 2002.

Lewis, Tom. *Empire of the Air: The Men Who Made Radio.* New York: Edward Burlingame Books, 1991.

———. "'A Godlike Presence': The Impact of Radio on the 1920s and 1930s." In "Communication in History: The Key to Understanding," special issue, *Magazine of History* 6, no. 4 (Spring 1992): 26–33.

Leyva, Gustavo, et al., eds. *Independencia y revolución: Pasado, presente y futuro.* Mexico City: Universidad Autónoma Metropolitana / Fondo de Cultura Económica, 2010.

Leyva, Juan. *Política educativa y comunicación social: La radio en México, 1940–1946.* Mexico City: Universidad Nacional Autónoma de México, 1992.

Lieuwen, Edwin. *Mexican Militarism: The Rise and Fall of the Revolutionary Army, 1910–1940.* Albuquerque: University of New Mexico Press, 1968.

Limantour, José Ives. "El capital extranjero." In *Mexico en el siglo XX, 1900–1913: Textos y documentos,* edited by Mario Contreras and Jesús Tamayo, 1:171. Mexico City: Universidad Nacional Autónoma de México.

Lornell, Kip. *Exploring American Folk Music: Ethnic, Grassroots, and Regional Traditions in the United States.* Oxford MS: University of Mississippi Press, 2012.

Los presidentes de México ante la nación: Informes, manifestos y documentos de 1821 a 1966. 5 vols. Mexico City: Cámara de Diputados, 1966.

Loyo Camacho, Martha Beatriz. *Joaquín Amaro y el proceso de institucionalización del ejército mexicano, 1917–1930.* Mexico City: Fondo de Cultura Económica, 2003.

Luz Ruelas, Ana. *México y Estados Unidos en la revolución mundial de las telecomunicaciones.* Mexico City: Universidad Autónoma de Sinaloa / Universidad Nacional Autónoma de México, 1996.

Macías Richard, Carlos. "El territorio de Quintana Roo: Tentativas de colonización y control militar en la selva maya (1888–1902)." *Historia Mexicana* 49, no. 1 (July–September 1999): 5–54.

———. *Nueva Frontera mexicana: Milicia, burocracia y ocupación territorial en Quintana Roo.* Chetumal: Consejo Nacional de Ciencia y Tecnología / Universidad de Quintana Roo, 1997.

MacLachlan, Colin M. *Anarchism and the Mexican Revolution: The Political Trials of Ricardo Flores Magón in the United States.* Berkeley: University of California Press, 1991.

Maples Arce, Manuel. *Las semillas del tiempo: Obra poética, 1919–1980.* Mexico City: Fondo de Cultura Económica, 1981.

Maples Arce, Manuel, et al. *El Estridentismo antología*. Mexico City: Difusión Cultural / Universidad Nacional Autónoma de México, 1983.

Martínez, Miranda, Elio Agustín, and María de la Paz Ramos Lara. "Funciones de los ingenieros inspectores al comienzo de las obras del complejo hidroeléctrico de Necaxa." *Historia Mexicana* 56, no. 1 (July–September 2006): 231–86.

Matthews, Michael. *The Civilizing Machine: A Cultural History of Mexican Railroads, 1876–1910*. Lincoln: University of Nebraska Press, 2013.

———. "*De Viaje*: Elite Views of Modernity and the Porfirian Railway Boom." *Mexican Studies / Estudios Mexicanos* 26, no. 2 (Summer 2010): 251–89.

Matute, Álvaro, ed. *Contraespionaje político y sucesión presidencial: Correspondencias de Trinidad W. Flores sobre la primera campaña electoral de Álvaro Obregón, 1919–1920*. Mexico City: Universidad Nacional Autónoma de México, 1985.

Mazzotto, Domencio. *Wireless Telegraphy and Telephony*. Translated by Selimo Romeo Bottone. London: Whittaker, 1906.

McCaa, Robert. *Missing Millions: The Human Cost of the Mexican Revolution*, University of Minnesota Population Center. http://www.hist.umn.edu/~rmccaa /missmill/mxrev.htm. Accessed April 10, 2014.

McCann, Bryan. "Carlos Lacerda: The Rise and Fall of a Middle-Class Populist in 1950s Brazil." *Hispanic American Historical Review* 83, no. 4 (Winter 2003): 661–96.

———. *Hello, Hello Brazil: Popular Music in the Making of Modern Brazil*. Durham NC: Duke University Press, 2004.

McCreery, David. "Wireless Empire: The United States and Radio Communications in Central America and the Caribbean, 1904–1925." *South Eastern Latin Americanist* 37 (Summer 1993): 23–41.

Medin, Tzvi. *El minimato presidencial: Historia política del Maximato, 1928–1935*. Mexico City: Ediciones Era, 1982.

Medina Ávila, Virginia and Gilberto Vargas Arana. *Nuestra es la voz, de todos la palabra: Historia de la radiodifusión mexicana, 1921–2010*. Mexico City: Universidad Nacional Autónoma de México, Facultad de Estudios Superiores Acatlán, 2011.

Mejía Barquera, Fernando. "Historia mínima de la radio mexicana (1920–1996)." *Revista de Comunicación y Cultura* 1, no. 1 (March–May 2007). http://web .upaep.mx/revistaeyc/radiomexicana.pdf. Accessed February 17, 2010.

———. *La industria de la radio y televisión y la política del estado mexicano (1920–1960)*. Mexico City: Fundación Manuel Buendía, 1989.

Mejía Prieto, Jorge. *Historia de la radio y la t.v. en México*. Mexico City: Editores Asociados, 1972.

Melgar, Arturo. "El desarrollo de la radiodifusión en México." *El Telegrafista* 2, no. 8 (February 1954): 25–26.

Memoria por la Secretaría de Comunicaciones y Obras Públicas. Mexico City, 1898–1938.

Memoria por la Secretaría de Guerra y Marina. Mexico City, 1898–1938.

Méndez Docurro, Eugenio. *Gral. Guillermo Garza Ramos y Trillo: Ejemplo de honor, lealtad y patriotismo*. Mexico City: IMC / CIME, 1994.

Merchán Escalante, Carlos A. *Telecomunicaciones*. Mexico City: Secretaría de Comunicaciones y Transportes, 1988.

The Mexican Constitution of 1917 Compared with the Constitution of 1857. Translated and arranged by H. N. Branch. Philadelphia: Annals of the American Academy of Political and Social Science, 1917.

Meyer, Lorenzo. *México y los Estados Unidos en el conflicto petróleo, 1917–1942*. Mexico City: Colegio de México, 1972.

———. "Un tema añejo siempre actual: El centro y las regiones en la historia mexicana." In *Descentralización y democracia en México*, edited by Blanca Torres. Mexico City: Colegio de México, 1986: 23–33.

Meyer, Michael C., and William H. Beezley, eds. *The Oxford History of Mexico*. Oxford: Oxford University Press, 2000.

Miller, Tom. *On the Border*. New York: Harper & Row, 1981.

Miñano Garcia, Max. H. *La educación rural en México*. Mexico City: Ediciones de la Secretaría de Educación Pública, 1945.

Miquel, Ángel. *Disolvencias: Literatura, cine y radio en México (1900–1950)*. Mexico City: Fondo de Cultura Económica, 2005.

Mora-Torres, Juan. *The Making of the Mexican Border: The State, Capitalism, and Society in Nuevo León, 1848–1910*. Austin: University of Texas Press, 2001.

Moreno, Julio. *Yankee Don't Go Home!: Mexican Nationalism, American Business Culture, and the Shaping of Modern Mexico, 1920–1950*. Chapel Hill: University of North Carolina Press, 2003.

Moreno Rivas, Yolanda. *Rostros del nacionalismo en la música Mexicana: Un ensayo interpretación*. Mexico City: Escuela Nacional de Música, Universidad Nacional Autónoma de México, 1993.

Mota Martínez, Fernando, and María Esther Núñez Herrera. *Locutores en acción: Vida y hazañas de quienes hicieron la radio Mexicana*. Mexico City: Times Editores, 1998.

Navarro, Aaron W. *Political Intelligence and the Creation of Modern Mexico, 1938–1954*. University Park: Pennsylvania State University Press, 2010.

Nemerov, Howard. *Figures of Thought: Speculations on the Meaning of Poetry and Other Essays*. Boston: David R. Godine, 1978.

Neulander, Joelle. *Programming National Identity: The Culture of Radio in 1930s France*. Baton Rouge: Louisiana State University Press, 2009.

Niblo, Stephen R. *War, Diplomacy, and Development: The United States and Mexico, 1938–1954*. Wilmington DE: SR Books, 1995.

Nickles, David Paull. *Under the Wire: How the Telegraph Changed Diplomacy*. Cambridge MA: Harvard University Press, 2003.

Norris, Renfro Cole. "A History of *La Hora Nacional:* Government Broadcasting via Privately Owned Radio Stations in México." PhD dissertation, University of Michigan, 1963.

Novelo, Victoriano. *Yucatecos en Cuba: Etnografía de una migración.* Mexico City: Publicaciones de la Casa Chata, 2009.

Novo, Salvador. *La vida en México en el período presidencial de Lázaro Cárdenas.* Mexico City: Consejo Nacional para la Cultura y las Artes, 1994.

Noyola, Leopoldo. *La raza de la herba: Historia de telégrafos Morse en México.* 2nd ed. Puebla: Benemérita Universidad Autónoma de Puebla, 2004.

Obregón, Álvaro. *Ocho mil kilómetros en campaña.* Mexico City: Fondo de Cultura Económica, 1959.

Official List of Radiotelegraph Stations Open for International Traffic. 2nd ed. Berne: International Telegraph Bureau, 1911.

Ornelas Herrera, Roberto. "La radiodifusión mexicana a principios del siglo XX: Las comunicaciones inalámbricas en México 1900–1924." Tesis de licenciatura, Universidad Nacional Autónoma de México, 1998.

———. "Radio y contidianidad en México (1900–1930)." In *Historia de la vida cotidiana en México.* Vol. 5, *Siglo XX: Campo y ciudad,* edited by Aurelio de los Reyes, 127–69. Mexico City: Colegio de México / Fondo de Cultura Económica, 2006.

Ortiz García, José Luis. *La guerra de las ondas: Un libro que desmiente la historia "oficial" de la radio mexicana.* Mexico City: Planeta Mexicana, 1992.

———. *Radio entre dos reinos: La increíble historia de la radiodifusora mexicana más potente del mundo en los años 30.* Mexico City: Vergara, 1997.

Papers relating to the Foreign Relations of the United States with the Annual Message of the President, Transmitted to Congress, December 3, 1906. Pt. 2. Washington DC: Government Printing Press, 1909.

Paz Salinas, María Emilia. *Strategy, Security, and Spies: Mexico and the U.S. as Allies in World War II.* University Park: Pennsylvania State Press, 1997.

Peña y Peña, Alvaro. *Territorio de Quintana Roo.* Mexico City: Secretaría Educación Pública, 1970.

Pérez Montfort, Ricardo. "'Esa no, porque me hiere': Semblanza superficial de treinta años de radio en Mexico, 1925–1955." In *Avatares del nacionalismo cultural: Cinco ensayos,* 91–115. Mexico City: Centro de Investigación y Docencia en Humanidades de Morelos / Centro de Investigaciones y Estudios Antropología Social, 2000.

Peterkin, Allan. *One Thousand Beards.* Vancouver: Arsenal Pulp Press, 2001.

Peterson, Adrian M. "Early Wireless Stations in the Philippines." *Wavescan.* http://www.ontheshortwaves.com/Wavescan/wavescan090830.html. Accessed February 21, 2012.

Plasencia de la Parra, Enrique. *Historia y organización de las fuerzas armadas en México, 1917–1937*. Mexico City: Universidad Nacional Autónoma de México, 2010.

———. *Personajes y escenarios de la Rebelión Delahuertista, 1923–1924*. Mexico City: Universidad Nacional Autónoma de México, 1998.

Popkin, Jeremy D. *Revolutionary News: The Press in France, 1789–1799*. Durham NC: Duke University Press, 1999.

Quirk, Robert E. *An Affair of Honor: Woodrow Wilson and the Occupation of Veracruz*. Lexington: University of Kentucky Press, 1962.

———. *The Mexican Revolution, 1914–1915*. New York: Citadel Press, 1963.

Radovsky, M. *Alexander Popov: Inventor of Radio*. Translated by G. Yankosky. Honolulu: University Press of the Pacific, 2001.

Raines, Rebecca Robbins. *Getting the Message Through: A Branch History of the U.S. Army Signal Corps*. Washington DC: Center of Military History, U.S. Army, 1996.

Rashkin, Elissa J. *The Stridentist Movement in Mexico: The Avant-Garde and Cultural Change in the 1920s*. Lanham MD: Lexington Books, 2009.

Reitz, Deneys. *Commando: A Boer Journal of the Boer War*. Pretoria, ZA: Cru-Guru, 2008.

Resler, Ansel Harlan. "The Impact of John R. Brinkley on Broadcasting in the United States." PhD dissertation, Northwestern University, 1958.

Richmond, Douglas W. *Venustiano Carranza's Nationalist Struggle, 1893–1920*. Lincoln: University of Nebraska Press, 1984.

Robles, Sonia. "Shaping *México Lindo*: Radio, Music, and Gender in Greater Mexico, 1923–1946." PhD dissertation, Michigan State University, 2012.

Rodríguez, Arnulfo. "Y llegó la comunicación sin cables: La primera transmission de radiotelefonía en México." *Relatos e Historias México* 35 (July 2011): 80–83.

Rodríguez, Clemencia. *Citizens' Media against Armed Conflict: Disrupting Violence in Colombia*. Minneapolis: University of Minnesota Press, 2011.

Rodriguez, Julia. *Civilizing Argentina: Science, Medicine, and the Modern State*. Chapel Hill: University of North Carolina Press, 2006.

Rolland, Modesto C. *Informe sobre el Distrito Norte de la Baja California de la Baja California*. Mexicali: Universidad Autónoma de Baja California, 1993.

———. *A Reconstructive Policy in Mexico*. New York: Latin-American News Association, 1917.

Rolle, Andrew F. "Futile Filibustering in Baja California." *Pacific Historical Review* 20, no. 2 (May 1951): 159–66.

Romero Aceves, Ricardo. *Baja California: Histórica y legendaria*. Mexico City: Costa-Amic Editores, 1983.

Romo Gil, María Cristina. *Introducción al conocimiento y práctica de la radio*. Mexico City: Editorial Diana, 1987.

Rosales Vargas, Manuel, and Virginia Licona Peña, eds. *Historia de las telecomunicaciones*. Mexico City: Telecomm / Telégrafos, 1999.

Rosenberg, Emily S. *Spreading the American Dream: American Economic and Cultural Expansion, 1890–1945.* New York: Hill & Wang, 1982.

Rubenstein, Anne. "Mass Media and Popular Culture in the Postrevolutionary Era." In *The Oxford History of Mexico,* edited by Michael C. Meyer and William H. Beezley, 637–70. New York: Oxford University Press, 2000.

Rudel, Anthony. *Hello, Everybody! The Dawn of American Radio.* Orlando: Harcourt, 2008.

Rugeley, Terry. *Rebellion Now and Forever: Mayas and Caste War Violence in Yucatán, 1800–1880.* Stanford CA: Stanford University Press, 2009.

———. *Yucatán's Maya Peasantry and the Origins of the Caste War, 1800–1847.* Austin: University of Texas Press, 1996.

Ruíz, Ramón Eduardo. *The Great Rebellion, 1905–1924.* New York: W. W. Norton, 1980.

Sakar, Tapan K, et al. *History of Wireless.* Hoboken NJ: John Wiley & Sons, 2006.

Sánchez, George I. *Mexico: A Revolution by Education.* New York: Viking, 1936.

Sánchez Ruiz, Enrique E. *Orígenes de la radiodifusión en México: Desarrollo capitalista y el estado.* Mexico City: ITESO, 1984.

Sandos, James A. *Rebellion in the Borderlands: Anarchism and the Plan de San Diego, 1904–1924.* Norman: University of Oklahoma Press, 1992.

Santiago, Myrna I. *The Ecology of Oil: Environment, Labor, and the Mexican Revolution, 1900–1938.* Cambridge: Cambridge University Press, 2006.

Saragoza, Alex M. *The Monterrey Elite and the Mexican State, 1880–1940.* Austin: University of Texas Press, 1990.

Sarlo, Beatriz. *The Technical Imagination: Argentine Culture's Modern Dreams.* Translated by Xavier Callahan. Stanford CA: Stanford University Press, 2008.

Sarton, George, and John Christian Oersted. "The Foundation of Electromagnetism." *Isis* 10, no. 2 (June 1928): 435–44.

Satia, Priya. "War, Wireless, and Empire: Marconi and the British Warfare State, 1896–1903." *Technology and Culture* 51, no. 4 (October 2010): 829–54.

Schantz, Eric Michael. "All Night at the Owl: The Social and Political Relations of Mexicali's Red-Light District, 1913–1925." In "Border Cities and Culture," special issue, *Journal of the Southwest* 43, no. 4 (Winter 2001): 449–602.

Schruben, Francis W. "The Wizard of Milford: Dr. J. R. Brinkley and Brinkleyism." *Kansas History* 14, no. 4 (Winter 1991–92): 226–45.

Schuler, Friedrich. *Secret Wars and Secret Policies in the Americas, 1842–1929.* Albuquerque: University of New Mexico Press, 2010.

Schwoch, James. *The American Radio Industry and Its Latin American Activities, 1900–1939.* Urbana: University of Illinois Press, 1990.

Scott, James Brown, ed. *The International Conferences of American States, 1898–1928.* New York: Oxford University Press, 1931.

Scott, James C. *Seeing like a State: How Certain Schemes to Improve the Human Condition Fail.* New Haven CT: Yale University Press, 1997.

Secretaría de Comunicaciones y Obras Públicas. *Ley de Comunicaciones Eléctricas*. Mexico City: Talleres Gráficos de la Nación, 1926.

Secretaría de Educación Pública. *La educación pública en México: A través de los mensajes presidenciales desde la consumación de la independencia hasta nuestros días*. Mexico City: Publicaciones de la Secretaría de Educación Pública, 1926.

———. *Las misiones culturales en 1927: Las escuelas normales rurales*. Mexico City: Publicaciones de la Secretaría de Educación Pública, 1928.

Secretaría de Fomento Colonización e Industria. *Censo de 1900*. Mexico City: Oficina Tipografía de la Secretaría de Fomento, 1901.

Shoup, G. Stanley. "The Control of International Radio Communications." In "Radio," supplement, *Annals of the American Academy of Political and Social Science* 142 (March 1929): S95–S104.

Smith, Michael M. "Carrancista Propaganda and the Print Media in the United States: An Overview of Institutions." *Americas* 52, no. 2 (October 1995): 155–74.

Sosa Plata, Gabriel. *Las mil y una radios: Una historia, un análisis actual de la radiodifusión mexicana*. Mexico City: McGraw-Hill, 1997.

Spicer, Edward H. *Cycles of Conquest: The Impact of Spain, Mexico, and the United States on the Indians of the Southwest, 1933–1960*. Tucson: University of Arizona Press, 1967.

Sterling, Christopher H. *Military Communications: From Ancient Times to the 21st Century*. Santa Barbara CA: ABC-CLIO, 2008.

Stewart, Irwin. "Recent Radio Legislation." *American Political Science Review* 23, no. 2 (May 1929): 412–26.

Sullivan, Paul. *Unfinished Conversations: Mayas and Foreigners between Two Wars*. Berkeley: University of California Press, 1991.

Süsskind, Charles. "Hertz and the Technological Significance of Electromagnetic Waves." *Isis* 56, no. 3 (Autumn 1965): 342–45.

Tannenbaum, Frank. *The Mexican Agrarian Revolution*. New York: MacMillan, 1929.

Taylor, Lawrence D. "The Mining Boom in Baja California from 1850 to 1890 and the Emergence of Tijuana as a Border Town." In "Border Cities and Culture," special issue, *Journal of the Southwest* 43, no. 4 (Winter 2001): 463–92.

Tenerio-Trillo, Mauricio. *Mexico at the World's Fair: Crafting a Modern Nation*. Berkeley: University of California Press, 1996.

Terman, Frederick Emmons. *Fundamentals of Radio*. New York: McGraw-Hill, 1938.

Terrazas, Silvestre. *El verdadero Pancho Villa*. Mexico City: Ediciones Era, 1985.

Torres, Blanco, ed. *Descentralización y democracia en México*. Mexico City: Colegio de México, 1986.

Torres, Carlos Alberto, ed. *Education and Social Change in Latin America*. Albert Park AU: James Nichols, 1995.

Tovey, David G. "The Role of the Music Educator in Mexico's Cultural Missions." *Bulletin of the Council for Research in Music Education* 139 (Winter 1999): 3–4.

Tuchman, Barbara W. *The Zimmerman Telegram*. New York: Ballantine Books, 1958.

Ulloa, Berta. "La lucha armada (1911–1920)." In *Historia general de México, 757–817*. Mexico City: Colegio de México, 2000.

———. *Veracruz, capital de la nación (1914–1915)*. Mexico City: Colegio de México, 1986.

U.S. Army. *Regulations Governing Commercial Radio Service between Ship and Shore Stations*. Washington DC: Government Printing Press, 1914.

U.S. Congress. House. Committee on Merchant Marine, Radio, and Fisheries. *Regulation of American Broadcasting Companies Operating Across the International Border*. 73rd Congress, 2nd Session. March 22, 1934. Washington DC: Government Printing Press, 1934.

U.S. Congress. House. Committee on Merchant Marine, Radio, and Fisheries. *Remote Control Border Stations—H.R. 7800*. 73rd Congress, 2nd Session. February 15, 1934. Washington DC: Government Printing Press, 1934.

U.S. Congress. House. Letter from Acting Secretary of War Henry Breckinridge to the U.S. Secretary of State, 6786 H.doc. 1721/2. Reproduced for the 63rd Congress, 3rd Session. August 26, 1913. Washington DC: Government Printing Press, 1913.

U.S. Congress. Senate. Committee on Interstate Commerce. *Commission on Communications*. 71st Congress, 1st Session. Vol. 1, May 8, 1929 to June 7, 1929. Washington DC: Government Printing Press, 1930.

U.S. Navy. *List of Wireless Stations of the World: Including Shore Stations, Merchant Vessels, Revenue Cutters, and Vessels of the United States Navy*. October 1, 1910. Washington DC: Government Printing Office, 1910. http://earlyradiohistory.us/1910stat.htm. Accessed March 11, 2010.

Walker, Jesse. *Rebels on the Air: An Alternative History of Radio in America*. New York: New York University Press, 2001.

Waterbury, John I. "The International Preliminary Conference to Formulate Regulation Governing Wireless Telegraphy." *North American Review* 177, no. 564 (November 1903): 655–56.

Whittemore, Laurens E. "The Development of Radio." In "Radio," supplement, *Annals of the American Academy of Political and Social Science*. Vol. 142 (March 1929): S1–S7.

Wilson, Henry Lane. *Diplomatic Episodes in Mexico, Belgium, and Chile*. New York: Doubleday, Page, 1927.

Wilson, Woodrow. "Wilson's Special Message on the Tampico Affair, April 10, 1914, Washington DC." In *The Messages and Papers of Woodrow Wilson*. New York: Review of Reviews, 1924.

Winseck, Dwayne R., and Robert M. Pike. *Communications and Empire*. Durham NC: Duke University Press, 2007.

Womack, John, Jr. *Zapata and the Mexican Revolution*. New York: Vintage, 1970.

Wood, Andrew Grant. *Revolution in the Street: Women, Workers, and Urban Protest in Veracruz, 1870–1927*. Wilmington DE: SR Books, 2001.

Wu, Tim. *The Master Switch: The Rise and Fall of Information Empires*. New York: Alfred A. Knopf, 2011.

Wyllys, Rufus Kay. "The Republic of California, 1853–54." *Pacific Historical Review* 2, no. 2 (June 1953): 194–213.

Vaughn, Mary Kay. *Cultural Politics in Revolution: Teachers, Peasants, and Schools in Mexico, 1930–1940*. Tucson: University of Arizona Press, 1997.

——. *The State, Education, and Social Class in Mexico, 1990–1928*. DeKalb: Northern Illinois University Press, 1982.

Vaughn, Mary Kay, and Stephen E. Lewis, eds. *The Eagle and the Virgin: Nation and Cultural Revolution in Mexico, 1920–1940*. Durham NC: Duke University Press, 2006.

Velázquez Estrada, Rosalía. "La radiodifusión mexicana durante los gobiernos de Alvaro Obregón y Plutarco Elías Calles." Tesis de licenciatura, Universidad Nacional Autónoma de México, 1980.

——. "La radiodifusión mexicana: Encuentro con su espada (1923–1945)." In *Miradas sobre la nación liberal: 1848–1948: Proyectos, debates y desafíos*. Libro 2, *Formar e informar: la diversidad cultural*, 275–313. Mexico City: Universidad Nacional Autónoma de México, 2010.

Vianna, Hermano. *The Mystery of Samba: Popular Music and National Identity in Brazil*. Edited and translated by John Charles Chasteen. Chapel Hill: University of North Carolina Press, 1999.

Yang, Daqing. *Technology of Empire: Telecommunications and Japanese Expansion in Asia, 1883–1945*. Cambridge MA: Harvard University Press, 2010.

INDEX

radio conventions (cont.)
(1927), 128, 156–57; International
Radio Telegraph Convention,
Berlin (1903), 38; International
Radio Telegraph Convention,
Berlin (1906), 38, 53–54;
International Technical
Consultative Committee on
Radioelectronic Communications,
Copenhagen (1930), 149–50;
International Telegraphic and
Radio Telegraphic Conference,
Madrid (1932), 149–50, 156, 161;
International Wireless Convention,
London (1912), 58; North and
Central American Regional Radio
Conference, Mexico City (1933),
161–62; Pan-American Conference,
Santiago, Chile (1923), 123, 155
Radio Corporation of America (RCA),
109, 113, 141, 149, 152, 155, 163
Radiomex, 152
radiotelegraph stations: 1-Z, 179;
Acapulco, 62; Algodones, 86;
Alamos, 81, 102; Balbuena, 107,
148; Bahía Magdalena, 54, 85;
Bocochibampo, 41, 54; Cabo Haro,
32, 41, 54; Campeche, 37, 54–55, 69,
129; Cerritos, 39; Chapultepec, 1,
51, 67, 76, 92–95, 102, 122, 131, 133,
137, 151–52; Chihuahua, 64, 81, 102;
Coahuayana, 193; Córdoba, 83;
Cuernavaca, 68–70, 82; Durango,
70, 145; El Aguila, 70, 181; Ensenada,
86, 193; Fort Bliss, 80; Fort Sam
Houston, 80; Guadalajara, 82–83;
Guadalajara air base, 145; Guaymas,
193; Hermosillo, 63, 102, 193;
Isla María Madre, 39, 55, 64–65;
Ixtapalapa, 92, 95; Juárez, 79–80;
La Paz, 193; Lomas de Santa Fe, 151;

Loreto, 54; Mazatlán, 55; Mérida,
132, 134, 193; Mexicali, 86; Miramar,
54; Oaxaca, 193; Pachuca, 107; Palo
Alto, 151; Payo Obispo, 35–36, 54;
Puerto Morelos, 85; Salina Cruz,
85, 132, 134; Saltillo, 76; San José
del Cabo, 39, 55, 85; San Quintín,
54; Santa Rosalía, 31, 41, 51–52;
Sayville, 87; Tampico, 76, 81; Tapa
Chula, 193; Tecate, 86; Tijuana, 86;
Torreón, 51, 65, 81; Túxpan, 129;
Tuxtla Gutiérrez, 193; Veracruz, 26,
54–55, 76, 129–31, 134; Villahermosa,
82; Waters-Pierce, 70; Xalapa, 152;
Xcalak, 35–36, 85
Ramírez, de Aguilar, Fernández, 170
Ramírez, José Fernando, 121, 133
Regional Confederation of Mexican
Workers (CROM), 133, 179–81
Reuthe, Gustavo, 92–93, 113, 125
Reyes, Bernardo, 21, 23, 27, 50, 56
Reynoso, José, 121
Ríos, Pedro, 133
Robles, Hernegildo, 149
Rodríguez, Abelardo, 140–41, 162,
188–89, 201
Rodríguez, Luis F., 170, 174
Rodríguez Gutiérrez, Manuel, 84, 103
Rolland, Modesto C., 81, 86, 111, 115,
117, 193–94
Ross, María Luisa, 186–88
Rouaix, Pastor, 96
Ruíz Sandoval, Manuel, 159

Sáenz, Moisés, 171–72
Sánchez, Guadalupe, 131, 133
Sánchez Ayala, Fernando, 60
Sánchez Martínez, Luis, 42, 89, 91–92,
125
San Martino, gunboat, 21
Sansalvador, C. E., 170, 174

The Lawyer of the Church: Bishop Clemente de Jesús Munguía and the Clerical Response to the Liberal Revolution in Mexico
Pablo Mijangos y González

¡México, la patria! Propaganda and Production during World War II
Monica A. Rankin

Murder and Counterrevolution in Mexico: The Eyewitness Account of German Ambassador Paul von Hintze, 1912–1914
Edited and with an introduction by Friedrich E. Schuler

Deco Body, Deco City: Female Spectacle and Modernity in Mexico City, 1900–1939
Ageeth Sluis

Pistoleros and Popular Movements: The Politics of State Formation in Postrevolutionary Oaxaca
Benjamin T. Smith

Alcohol and Nationhood in Nineteenth-Century Mexico
Deborah Toner

To order or obtain more information on these or other University of Nebraska Press titles, visit nebraskapress.unl.edu.

9 780803 286788